Electromagnetic Waves 1

SCIENCES

Waves, Field Directors – Pierre-Noël Favennec, Frédérique de Fornel

Electromagnetism, Subject Head – Pierre-Noël Favennec

Electromagnetic Waves 1

Maxwell's Equations, Wave Propagation

Coordinated by

Pierre-Noël Favennec

iSTE

WILEY

First published 2020 in Great Britain and the United States by ISTE Ltd and John Wiley & Sons, Inc.

ISTE Ltd
27-37 St George's Road
London SW19 4EU
UK

www.iste.co.uk

John Wiley & Sons, Inc.
111 River Street
Hoboken, NJ 07030
USA

www.wiley.com

Library of Congress Control Number: 2020937434

British Library Cataloguing-in-Publication Data
A CIP record for this book is available from the British Library
ISBN 978-1-78945-006-4

ERC code:
PE2 Fundamental Constituents of Matter
PE2_6 Electromagnetism

MIX
Paper from responsible sources
FSC® C013604

Contents

Preface ix

Chapter 1. Maxwell's Equations 1
Ibrahima SAKHO

1.1. Maxwell's equations in a vacuum. 1
1.1.1. Electrostatics. 1
1.1.2. Magnetostatics. 17
1.1.3. Electromagnetic induction. 33
1.1.4. Maxwell's equations 54
1.2. Maxwell equations in material media 85
1.2.1. Electric field and potential in macroscopic dielectric media 86
1.2.2. Homogeneous linear dielectric media 95
1.2.3. Magnetic media 98
1.2.4. Maxwell equations in a polarized and magnetic medium. 111
1.3. References 117

Chapter 2. The Propagation of Optical and Radio Electromagnetic Waves 119
Hervé SIZUN

2.1. Introduction. 119
2.2. Maxwell's equations 121

2.2.1. Maxwell-Gauss equation. 121
2.2.2. Maxwell-Thompson equation. 122
2.2.3. Maxwell-Faraday equation 123
2.2.4. Maxwell-Ampère equation 123
2.3. Solving Maxwell's equations 124
2.4. Characteristics of electromagnetic waves 125
2.4.1. Propagation speed. 125
2.4.2. Wavelength and/or frequency. 126
2.4.3. The characteristic impedance of the propagation medium 127
2.4.4. Poynting vector 127
2.4.5. The refractive index 128
2.4.6. Polarization 129
2.4.7. Transpolarization 131
2.4.8. Different propagation paths 132
2.4.9. Fresnel zones 133
2.4.10. Fundamental properties of the propagation channel 134
2.5. Propagation modeling. 146
2.5.1. Tropospheric propagation 147
2.5.2. Propagation in rural, suburban and urban areas 172
2.5.3. Propagation within buildings 184
2.5.4. Broadband propagation 196
2.5.5. Ultra-wideband propagation. 200
2.6. The propagation of visible and infrared waves in the Earth's atmosphere. 207
2.6.1. Introduction 207
2.6.2. The propagation of light in the atmosphere. 208
2.6.3. The different models 214
2.6.4. Experimental results 222
2.6.5. Fog and mist. 225
2.6.6. Sandstorms. 226
2.6.7. Meteorological optical range 227
2.6.8. Applications 231
2.7. Conclusion 232
2.8. Recommendations ITU-R 233
2.9. References 233

Appendix 1 239

Appendix 2 243

Appendix 3 261

Appendix 4 269

Appendix 5 273

List of Acronyms and Constants 275

List of Authors 277

Index 279

Preface

Pierre-Noël Favennec
ArmorScience, Lannion, France

Any electric charge set in motion produces electromagnetic radiation which propagates in space. This property is the basis of radioelectric, or photonic radiation production, used in particular in radio, television and communication systems among others. Any system supplied with electricity, or any element provided with electric charge, emits electromagnetic radiation and generates an electric and/or magnetic field in its close, or even distant, vicinity which is known as an "electromagnetic field".

Before Maxwell's work, we understood physical reality in terms of material points. After it, we represented physical reality with continuous fields. The concept of a field finds its origin, and its name, in the idea of describing a physical phenomenon from an underlying medium, which would explain the physical properties of space (a field of forces for a field of wheat subjected to the wind). Following Maxwell's research, the fields acquired an autonomous existence and reached the status of physical beings in their own right, no longer describing "the place where" but "the thing that". This movement was largely supported by the development of the mathematical formalism of the fields, in terms of partial differential equations. This, with regard to electricity and magnetism, is the content of Maxwell's theory which he published in 1861.

Maxwell is one of the greatest scientists, who changed our view of the world. He made a decisive contribution to the unifying and synthetic vision of electricity and

For a color version of all figures in this book, see www.iste.co.uk/favennec/electromagnetic1.zip.

magnetism. He showed that two domains, that of electric charges and their interactions and that of currents and magnetism, were only two facets of the same problem. Synthesized by four equations combining in the same formalism their respective characteristic magnitudes. He stated these interactions in clear mathematical language: Maxwell's equations. The vision of a universe formed by particles was succeeded by a world governed by fields, acting from a distance.

The electromagnetic field is the set of vector fields $(\vec{E}, \vec{B})$. The properties of the electromagnetic field at a point in space are determined by the properties of the electric field $\vec{E}$ and the magnetic field $\vec{B}$ at a point. In physics, the term "field" refers to the situation where we are in the presence of a physical magnitude distributed in a given region of space. This magnitude has a value determined at each point in this space and at all times. Having an area of space where there is an electromagnetic field, means that at each point in this space, we have two vector variables $\vec{E}$ and $\vec{B}$.

Electromagnetic waves are produced by excited matter. The deexcitation of the excited source produces around it a periodic variation of the electromagnetic field which propagates gradually in the vacuum at the phase speed (or propagation speed) close to 300,000 km per second. Depending on their emission frequency domain, they have different names: radio waves for the lowest frequencies, infrared waves, visible optical waves, then ultraviolet, then for the highest frequencies, X-rays and gamma rays. The electromagnetic wave propagates: a variable electric field generates a variable magnetic field and conversely a variable magnetic field generates a variable electric field. The conjoint propagation of these variations in a region constitutes a continuous wave phenomenon, capable of propagating (across the vacuum at 300,000 kilometers per second), transporting energy without the need for material support.

Waves are vibrations that propagate from one place to another in space, in a material medium or in a vacuum. Electromagnetic vibrations (electromagnetic waves) are waves obeying the laws of electromagnetism. Mechanical vibrations (pendulum, acoustics, etc.) obey the laws of mechanics, but often these mechanical vibrations are in fact fundamentally electromagnetic, due to the electromagnetic interactions of atoms and molecules of materials; they are described by "approximate" laws according to movements following the laws of mechanics.

Characterizing or measuring an electromagnetic field is carried out via current or voltage measurements. The electromagnetic field located in one place is the set of vector fields ($\vec{E}$, $\vec{B}$). *Sensors or antennas* measure, at a given point, the currents or voltages resulting from the field from the different vector magnitudes $\vec{E}$ and $\vec{B}$. Subsequent processing can, if useful, select the different frequencies.

In our everyday life, the *environmental electromagnetic field* does not arise from a single source. There are fields of natural origin (the sun, galaxy, geomagnetism, etc.) and those of human origin (household materials, transport, telecommunications, energy supply, etc.). Each point on the planet is subjected to a fairly intense electromagnetic "bath" depending on its location. The drawing below, envisaged by Michel Urien, shows that we are all "willingly" bathed in these electromagnetic waves. Let us try to understand our *electromagnetic environment*!

Figure P.1. *A wave bath envisaged by Michel Urien*[1]

This referenced work, presented in two inseparable volumes, is essential for any student, engineer or researcher wishing to understand electromagnetism and all the technologies derived from it.

Volume 1 is oriented towards the basic phenomena explaining electromagnetism: the famous Maxwell equations – essential to know – then the propagation phenomena of electromagnetic waves. It only concerns non-ionizing radiation, which is radiation from waves whose energies are insufficient to ionize an atom, that is to say incapable of removing an electron from matter. This excludes all radiation with an energy greater than 12.4 eV, that is that generated by X-ray and gamma ray emitters. This work is made up of two chapters.

1 Source: www.armorscience.com.

In **Chapter 1, Ibrahima Sakho** presents the Maxwell equations as clearly as possible. These equations are essential to comprehensively approach electromagnetism and all its derived fields such as radioelectricity, photonics, geolocation, measurement, telecommunications, medical imagery, radio astronomy, etc.

In **Chapter 2, Hervé Sizun** describes the propagation phenomena of electromagnetic, radio and photonic waves. Many factors, often complex, must be taken into account to properly understand these propagation problems in free and sometimes confined spaces.

In **Volume 2, Jean-Pierre Blot**, expert in *radio antennas* of all configurations, directs his analysis towards antennas, essential elements for the detection of electromagnetic waves, their characterization and use. This volume is intended to describe what an effective antenna should be, according to various parameters and conditions of use. It does not address the detection problems specific to photonics. Photonics and these detection problems will be seen in a future publication of the "Waves" series.

Important appendices with essential information, presenting in particular mathematical tools, complete these two volumes.

References

Cartini, R. (1993). *Panorama encyclopédique des sciences*. Belin, Paris.

de Fornel, F., Favennec, P.-N. (eds) (2007). Mesures en électromagnétisme. *Revue RS série I2M*, 7(1–4).

Favennec, P.-N. (2008). *Mesures de l'exposition humaine aux champs radio-électriques – Environnement radioélectrique*. Techniques de l'Ingénieur, Saint-Denis.

Serres, M., Farouki, N. (1997). *Dictionnaire des sciences*. Flammarion, Paris.

1

Maxwell's Equations

Ibrahima SAKHO
Université de Thiès, Senegal

1.1. Maxwell's equations in a vacuum

Our aim is to introduce the fundamental equations of electromagnetism followed by the four Maxwell equations in vacuum. These equations express the local relationships between the electric $\vec{E}(\vec{r},t)$ and magnetic $\vec{B}(\vec{r},t)$ fields and their sources constituted by the free charge $\rho(\vec{r},t)$ and current $\vec{J}(\vec{r},t)$ densities. From an educational point of view, the formulation of Maxwell's equations in vacuum necessarily involves the study of the three fundamental parts of electromagnetism, which are electrostatic, magnetostatic and induction. The study of these three parts will establish Gauss's theorem, Ampère's theorem, Faraday's law and the local law of the magnetic field translating the conservation of magnetic flux. From this study we will deduce Maxwell's equations for their applications to the study of the propagation of electromagnetic waves in vacuum and antenna radiation.

1.1.1. *Electrostatics*[1]

1.1.1.1. *Coulomb's law: electrostatic field*

A charged particle, assumed to be a point particle, carries an electric charge denoted q. Two point charges q_1 and q_2, spaced r apart and placed in a vacuum, are

For a color version of all figures in this book, see www.iste.co.uk/favennec/electromagnetic1.zip.

1 (Annequin and Boutigny 1974; Bok and Hulin-Jung 1979; Bruneaux *et al.* 2002; Amzallag *et al.* 2006; Benson 2015; Sakho 2018).

in electrostatic interaction. According to the principle of interactions, the electrostatic force $\vec{f}_{1\to2}$ exerted by charge q_1 on charge q_2 is equal and opposite to the electrostatic force $\vec{f}_{2\to1}$ exerted by charge q_2 on charge q_1. According to *Coulomb's law*:

$$\vec{f}_{1\to2} = -\vec{f}_{2\to1} = \frac{kq_1q_2}{r^2}\vec{e}_r \tag{1.1}$$

In equation [1.1], k is the electric constant $k = \dfrac{1}{4\pi\varepsilon_0} = 9 \times 10^9$ SI and ε_0 denotes the dielectric permittivity of the vacuum: $\varepsilon_0 \approx 8.84 \times 10^{-12}\ \mathrm{F \cdot m^{-1}}$.

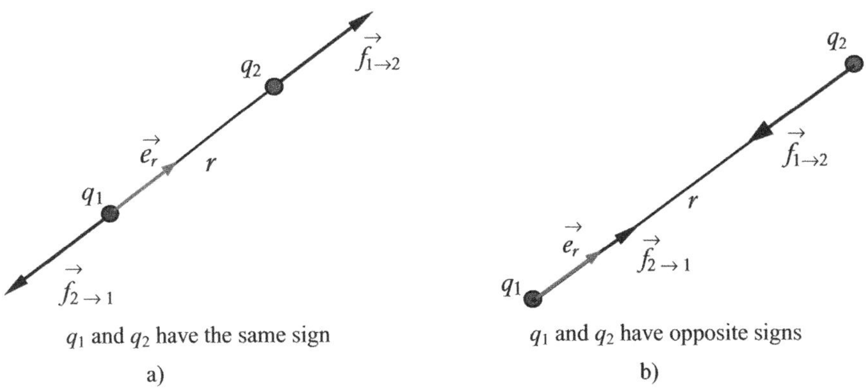

Figure 1.1. *Coulomb forces between two point and fixed charges q_1 and q_2*

The electrostatic forces $\vec{f}_{1\to2}$ and $\vec{f}_{2\to1}$ are repulsive if the charges q_1 and q_2 have the same sign and attractive if the two charges have opposite signs (Figure 1.1).

Now consider a test charge denoted q_0 fixed at a point O in domain $\mathcal{D}$ free of charges and currents. Let us place at a distance r from charge q_0 a charge q_1 fixed at point M ($OM = r$). The charge q_1 is subject to the electrostatic force $\vec{f}_{0\to1}$ from q_0. Let us remove the charge q_1 and place another fixed charge q_2 at the distance r from charge q_0. It is also subject to the electrostatic force $\vec{f}_{0\to2}$ from q_0. Let's repeat the experiment several times, taking care to leave in domain $\mathcal{D}$ only charge q_0 and a single charge q_i (i =1, 2, 3, …). Each of the point charges q_i is subject to an electrostatic force $\vec{f}_{0\to i}$ in domain $\mathcal{D}$ such that:

$$\vec{f}_{0\to 1} = \frac{kq_0q_1}{r^2}\vec{e}_r\,;\ \vec{f}_{0\to 2} = \frac{kq_0q_2}{r^2}\vec{e}_r\,;\ \vec{f}_{0\to 3} = \frac{kq_0q_3}{r^2}\vec{e}_r\ \ldots\ \vec{f}_{0\to i} = \frac{kq_0q_i}{r^2}\vec{e}_r \qquad [1.2]$$

Let's express the ratios $\vec{f}_{0\to i}/q_i$. Using [1.2], we obtain:

$$\frac{\vec{f}_{0\to 1}}{q_1} = \frac{\vec{f}_{0\to 2}}{q_2} = \frac{\vec{f}_{0\to 3}}{q_3} = \ldots\ldots \frac{\vec{f}_{0\to i}}{q_i} = \frac{kq_0}{r^2}\vec{e}_r \qquad [1.3]$$

The result [1.3] shows that the ratios $\vec{f}_{0\to i}/q_i$ are equal and depend only on the charge q_0 and the distance r between charge q_0 and charge q_i. In domain $\mathcal{D}$ there is therefore a vector field created by the charge q_0 known as the electrostatic field, denoted $\vec{E}_M$.

By definition, the electrostatic field $\vec{E}_M$ is equal to the ratios $\vec{f}_{0\to i}/q_i$, so according to [1.3] (with $q_0 = q$):

$$\vec{E}_M = \frac{kq_0}{r^2}\vec{e}_r \qquad [1.4]$$

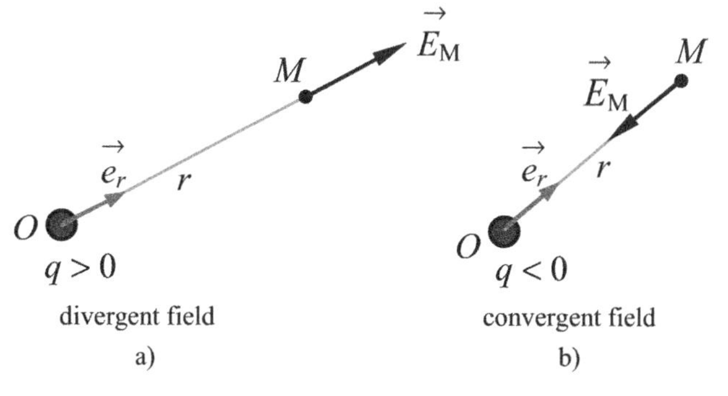

Figure 1.2. *Electrostatic field $\vec{E}_M$ created by a fixed point charge q*

The electrostatic field lines are oriented curves and tangent at each point to the electrostatic field. For a point charge at point O, the field lines are divergent if $q > 0$ and convergent if $q < 0$. On Figure 1.2, a single field line (in green) has been shown to coincide with the direction of the electrostatic field. The expression [1.4] shows that there is an electric monopole, the source of the electrostatic field $\vec{E}_M$. In addition, according to [1.4], the electrostatic field created by the point charge q at point M in space is a radial field in $1/r^2$. The electrostatic field $\vec{E}_M$ is divergent if $q > 0$ (Figure 1.2a) and convergent if $q < 0$ (Figure 1.2b). Furthermore, according to [1.3] and [1.4], $\vec{f}_{0\to i}/q_i = \vec{E}_M$. Thus, a point charge q in an area of space where an electrostatic field $\vec{E}$ prevails is subjected to the Coulomb force:

$$\vec{f} = q\vec{E} \qquad [1.5]$$

1.1.1.2. *Electrostatic field circulation: electrostatic potential*

Let us determine the circulation C of the electrostatic field according to direction OM between point M at a distance of r from point O and point M' at a distance of distance r' from the same point O ($r' > r$). By definition:

$$C = \int \vec{E} \cdot \vec{dr} \qquad [1.6]$$

Using [1.4], we obtain the following, according to [1.6]:

$$C = kq\int \frac{dr}{r^2} = -kq\left[\frac{1}{r}\right]_r^\infty = \left[\frac{kq}{r} - \frac{kq}{r_\infty}\right]$$

By definition, the circulation C of the electrostatic field between points M and M' is equal to the potential difference between these two points, i.e:

$$C = V_M - V_{M'} = \left[\frac{kq}{r} - \frac{kq}{r_\infty}\right] \qquad [1.7]$$

By definition, the electrostatic potential created by a point charge q fixed at point M in space characterizes the electrical state of this point. It is given by the expression:

$$V_M = \frac{kq}{r} + K \qquad [1.8]$$

In equation [1.8], K is a constant defining the origin of the electrostatic potential (generally, we choose $K = 0$ to infinity).

Is there an equation between an electrostatic field and electrostatic potential? To answer this question, let us express the gradient of the electrostatic potential V_M. In spherical coordinates, the radial component of the gradient of a scalar function f is written:

$$\vec{\nabla} f = \frac{\partial f}{\partial r}\vec{e_r} + \frac{1}{r}\frac{\partial f}{\partial \theta}\vec{e_\theta} + \frac{1}{r\sin\theta}\frac{\partial f}{\partial \varphi}\vec{e_\varphi} \qquad [1.9]$$

Using [1.8] and [1.9], we obtain the following, considering [1.4]:

$$\vec{\nabla} V \cdot \vec{e_r} = \frac{dV}{dr}\vec{e_r} = kq\frac{d}{dr}\left(\frac{d}{r}\right)\vec{e_r} = -\frac{kq}{r^2}\vec{e_r} = -\vec{E}$$

The result above allows us to write the important local equation linking the electrostatic field and the electrostatic potential:

$$\vec{E} = -\vec{\nabla} V \qquad [1.10]$$

Knowing the expression of the electrostatic field, we deduce the potential V from the local law [1.10] and vice versa.

Charles-Augustin Coulomb was a French officer, engineer and physicist. He is best known for his many experiments in electrostatics using a torsional balance to determine the force exerted between two electrical charges. He became famous following the publication of Coulomb's law in 1785, which forms the basis of electrostatics.

Box 1.1. *Coulomb (1736–1806)*

Using equation [1.6], we express the circulation of the electrostatic field on a closed contour (C) (Figure 1.3). We obtain:

$$C = \oint \vec{E} \cdot \vec{dl} = \int_M^M \vec{E} \cdot \vec{dl} = V(M) - V(M) = 0 \qquad [1.11]$$

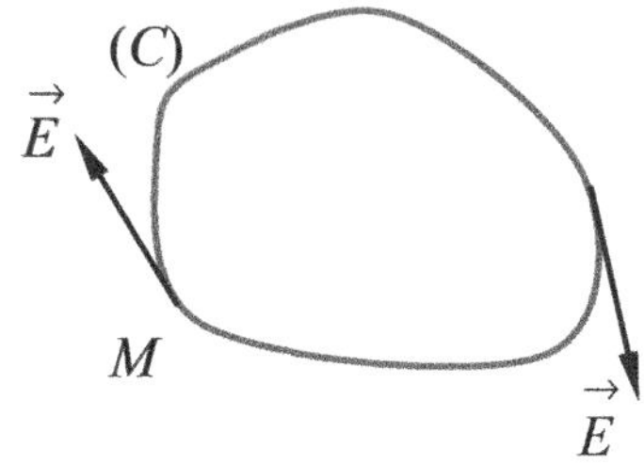

Figure 1.3. *Circulation of the electrostatic field about a closed contour (C)*

The circulation of the electrostatic field about a closed contour is therefore *zero*.

Let (*S*) be a surface based on contour (*C*). Using *Stokes' theorem*, we obtain the following, according to [1.11]:

$$\oint \vec{E} \cdot \vec{dl} = \iint_S (\vec{\nabla} \wedge \vec{E}) \cdot \vec{dS} = 0$$

With this equation being verified regardless of the contour or surface (*S*), we establish one of the local laws satisfied by the electrostatic field:

$$\vec{\nabla} \wedge \vec{E} = \vec{0} \qquad [1.12]$$

George Gabriel Stokes was a British physicist and mathematician. He is especially famous for having established Stokes' theorem in 1854, which is one of the fundamental theorems of integral transformation linking the rotational of a vector field to the circulation of the same field along a closed contour. Stokes' theorem is widely used in electrostatics and magnetostatics.

Box 1.2. *Stokes (1819–1803)*

1.1.1.3. *Electrostatic field and potential of a continuous charge distribution*

For a continuous charge distribution, we define an infinitesimal charge dq associated with a linear λ, surface σ or volume ρ charge:

– linear distribution over a wire with a length dl: $dq = \lambda dl$; λ in $\mathrm{C \cdot m^{-1}}$;

– surface distribution over a surface dS: $dq = \sigma dS$; σ in $\mathrm{C \cdot m^{-2}}$;

– volume distribution in a volume dV: $dq = \rho\, dV$; ρ in $\mathrm{C \cdot m^{-3}}$.

Let us consider various continuous charge distributions (Figure 1.4).

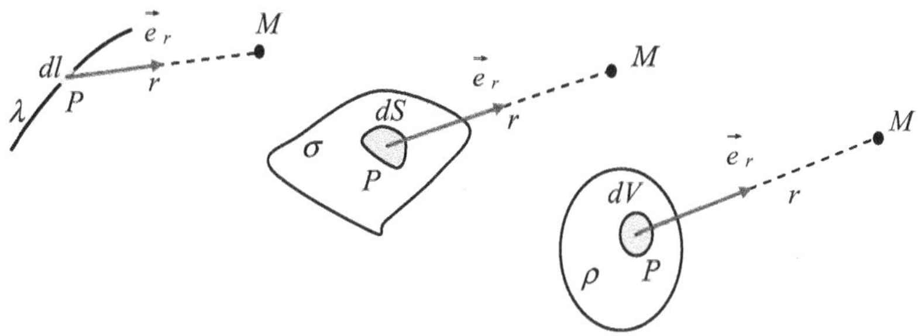

Figure 1.4. *Continuous charge distributions*

For these charge distributions, the electrostatic field $\vec{E}$ and the electrostatic potential V are given by the following expressions:

– linear distribution l (Figure 1.4a):

$$\vec{E} = \int_{AB} \frac{\lambda\, dl}{4\pi\varepsilon_0\, r^2} \vec{e}_r \quad [1.13a]$$

$$V = \int_{AB} \frac{\lambda\, dl}{4\pi\varepsilon_0\, r} \quad [1.13b]$$

– surface distribution S (Figure 1.4b):

$$\vec{E} = \iint_S \frac{\sigma\, dS}{4\pi\varepsilon_0\, r^2} \vec{e}_r \quad [1.14a]$$

$$V = \iint_S \frac{\sigma\, dS}{4\pi\varepsilon_0\, r} \quad [1.14b]$$

– volume distribution V (Figure 1.4c):

$$\vec{E} = \iiint_V \frac{\rho\, dV}{4\pi\varepsilon_0\, r^2} \vec{e}_r \quad [1.15a]$$

$$V = \iiint_V \frac{\rho\, dV}{4\pi\varepsilon_0\, r} \quad [1.15b]$$

1.1.1.4. *Flux and divergence of the electrostatic field: Gauss's law*

Let us consider the lines of an electrostatic field about a closed contour (*C*) (Figure 1.5).

By definition, the flux **Φ** of the electrostatic field across a surface *dS* is given by the following expression:

$$\Phi = \iint_S \vec{E}\cdot\overrightarrow{dS} = \iint_S \vec{E}\cdot\vec{n}\,dS \qquad [1.16]$$

In expression [1.16], $\vec{n}$ is the unit vector normal to surface *dS*.

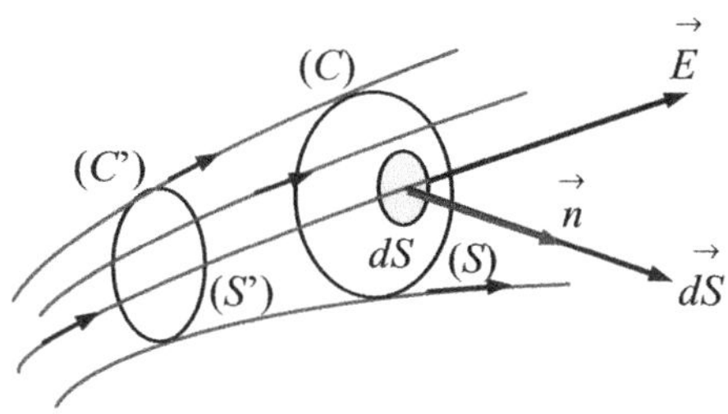

Figure 1.5. *Electrostatic field lines*

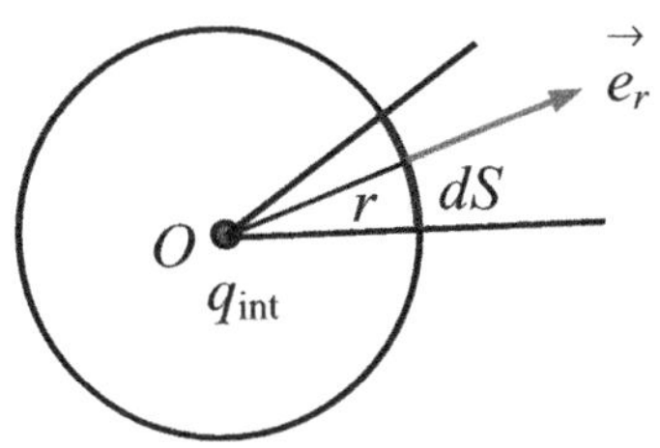

Figure 1.6. *Charge q_{int} at the center O of a sphere with radius r*

A spherical domain in space has a volume of *V*. We place, at the center *O* of the sphere with radius *r*, an internal charge q_{int} (Figure 1.6). This charge creates, at point *M* located on the surface and the electrostatic field deduced from [1.6]:

$$\vec{E} = \frac{kq_{\text{int}}}{r^2}\vec{e}_r \qquad [1.17]$$

The elementary flux $d\Phi$ of the field [1.17] across the element surface dS is equal to:

$$d\Phi = \vec{E} \cdot \overrightarrow{dS} = kq_{\text{int}} \frac{\overrightarrow{dS} \cdot \vec{e}_r}{r^2} \qquad [1.18]$$

By definition, the elementary solid angle $d\Omega$ is given by equation:

$$d\Omega = \frac{\overrightarrow{dS} \cdot \vec{e}_r}{r^2} \qquad [1.19]$$

Using expression [1.19], the flux [1.18] is written as:

$$d\Phi = k\, q_{\text{int}} d\Omega$$

Knowing that $k = 1/4\pi\varepsilon_0$, the total flux across the surface S is:

$$\Phi = \frac{q_{\text{int}}}{4\pi\varepsilon_0}\Omega \qquad [1.20]$$

In the whole space, the solid angle $\Omega = 4\pi$. The flux of the electrostatic field created by the charge q_{int} across a closed surface (S) is therefore given by:

$$\Phi = \iint_{(S)} \vec{E} \cdot \overrightarrow{dS} = \frac{q_{\text{int}}}{\varepsilon_0} \qquad [1.21]$$

The result [1.21] is generalized to a distribution of N point charges. Therefore, the flux of the electrostatic field created by a charge distribution across a closed surface (S) is equal to the quotient of the algebraic sum $Q_{\text{int}} = \sum_{i=1}^{N} q_i$ of the charges contained in volume V by the dielectric permittivity of the vacuum ε_0, i.e.:

$$\Phi = \iint_{(S)} \vec{E} \cdot \overrightarrow{dS} = \sum_i \frac{q_i}{\varepsilon_0} = \frac{Q_{\text{int}}}{\varepsilon_0} \qquad [1.22]$$

Expression [1.22] transcribes *Gauss's theorem*.

Johann Carl Friedrich Gauss was a German astronomer, physicist and mathematician. He formulated the mathematical theory of the magnetometer in his work *General Theory of Terrestrial Magnetism* (originally titled *Allgemeine Theorie des Erdmagnetismus* in German) published in 1839. Gauss's law for electrostatic fields expresses that a positive electrical charge (monopole) creates a divergent field. In physics, Gauss is especially famous for his theorem (Gauss's theorem) formulated in 1840, allowing the simple calculation of the electrostatic field module created by a charge distribution when the system has symmetry (cylindrical or spherical).

Box 1.3. *Gauss (1777–1855)*

Now let us consider a continuous charge distribution in a domain with volume V. The elementary charge $dq = \rho dV$, ρ denoting the charge density. The total charge is equal to $Q_{\text{int}} = \iiint_V \rho\, dV$. The elementary flux of the electrostatic field created by the continuous charge distribution across a closed surface (S) of volume V is written, according to [1.22]:

$$d\Phi = \vec{E}\cdot\overrightarrow{dS} = \frac{dq}{\varepsilon_0} = \frac{\rho\, dV}{\varepsilon_0}$$

The total flux total across a surface S is therefore equal to:

$$\Phi = \iint_{(S)} \vec{E}\cdot\overrightarrow{dS} = \iiint_{(V)} \frac{\rho\, dV}{\varepsilon_0} \qquad [1.23]$$

According to the *divergence theorem or Ostrogradsky's theorem* (also known as the Green-Ostrogradsky theorem in French):

$$\iint_{(S)} \vec{E}\cdot\overrightarrow{dS} = \iiint_{(V)} (\vec{\nabla}\cdot\vec{E})dV \qquad [1.24]$$

Using [1.23] and [1.24], we obtain:

$$\iiint_{(V)} \left(\vec{\nabla}\cdot\vec{E} - \frac{\rho}{\varepsilon_0}\right) dV = 0$$

The equality above is true regardless of the volume V, hence:

$$\vec{\nabla}\cdot\vec{E} = \frac{\rho}{\varepsilon_0} \qquad [1.25]$$

Equation [1.25] transcribes *the local form of Gauss's theorem* and is one of Maxwell's fundamental equations in a vacuum.

George Green was a British physicist. He is especially famous for *An Essay on the Application of Mathematical Analysis to the Theories of Electricity and Magnetism*, published in 1828. In this essay, he introduces several concepts, such as electric potential.

Mikhail Vasilyevich Ostrogradsky was a Russian mathematician and physicist of Ukrainian origin. Ostrogradsky is especially famous for having established, independently of Green, the divergence theorem, also known as Ostrogradsky's theorem. This theorem, demonstrated by Green in 1826, is one of the fundamental theorems of integral transformation linking the divergence of a vector field to the flow of the same field through any closed surface.

Box 1.4. *Green (1793–1841) and Ostrogradsky (1801–1861)*

1.1.1.5. *Poisson and Laplace equations*

Using the local law [1.10] linking the electrostatic potential and field, the local form [1.25] of Gauss's law gives:

$$\vec{\nabla}\cdot\vec{E} = -\vec{\nabla}\cdot(\vec{\nabla}V) = \frac{\rho}{\varepsilon_0} \qquad [1.26]$$

Let's take into account the following property satisfied by a scalar function *f*:

$$\vec{\nabla}\cdot(\vec{\nabla}f) = \nabla^2 f = \Delta f$$

Using this property, equation [1.26] gives:

$$\Delta V + \frac{\rho}{\varepsilon_0} = 0 \qquad [1.27]$$

The result [1.27] expresses *Poisson's equation*. In particular, in a vacuum, the volume charge density $\rho = 0$. Poisson's equation leads to *Laplace's equation*:

$\Delta V = 0$.

Pierre Simon Laplace was a physicist, mathematician astronomer, and French politician. He is famous in physics for the Laplace equation, which expresses that the Laplacian of the electrostatic potential is zero in a vacuum. In addition, between 1982 and 1784, the magnetic force on a current-carrying wire was named Laplace force.

Siméon Denis Poisson was a French mathematician, physicist and surveyor. His contributions in physics relate to electricity and magnetism. Drawing on Laplace's equation, in 1813, Poisson published the differential equation satisfied by all electrostatic potential. Based on the flux of a vector field, the local Poisson equation was formulated on a macroscopic scale by Gauss in 1840 (Gauss's theorem).

Box 1.5. *Laplace (1749–1827) and Poisson (1781–1840)*

1.1.1.6. *Field and potential of an electrostatic dipole*

An electrostatic dipole is composed of two opposite point charges $-q$ and $+q$ in two points A and B in space (Figure 1.7a). By definition, the dipolar moment $\vec{p}$ is:

$$\vec{p} = q\overrightarrow{AB} = qa\vec{u} \qquad [1.28]$$

The electric dipole moment is orientated from the negative charge towards the positive charge since $q > 0$. The units for electric dipole moment are coulombs-meters (C · m). However, a commonly used unit in for the electric dipole moment is the Debye (Dy) with:

$$1 \text{ Dy} = (1/3) \times 10^{-29} \text{C} \cdot \text{m}$$

Furthermore, the appearance of field lines and equipotential surfaces (sets of points brought to the same potential) of an electrostatic dipole are indicated in Figure 1.7b.

We express the scalar potential $V(r,\theta)$ and the field $\vec{E}(r,\theta)$ created by the electrostatic dipole AB (Figure 1.7a) at point M, ($r >> a$) far from point O. First, we determine the potential and, from this, the field $\vec{E}(r,\theta)$ using the local law [1.10].

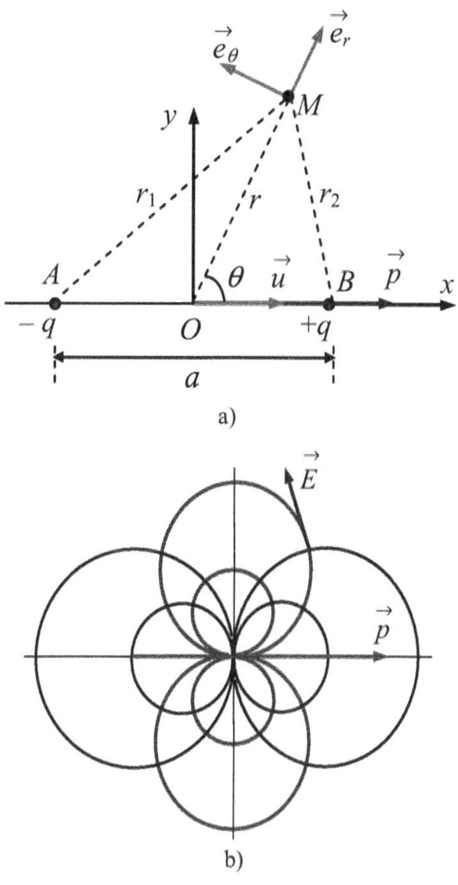

Figure 1.7. *a) Electrostatic dipole; b) field lines (in green) and equipotential surfaces (in blue) of an electrostatic dipole*

According to the superposition principle, the potential created by the charges $-q$ and $+q$ in M (Figure 1.7a) is given by the expression:

$$V(r_1, r_2) = -k\frac{q}{r_1} + k\frac{q}{r_2} = kq\left(\frac{1}{r_2} - \frac{1}{r_1}\right) \qquad [1.29]$$

Considering Figure 1.7a, we express the radii r_1 and r_2 as a function of the variables r and θ. We then obtain the following expressions:

$$\vec{r_1} = \vec{r} + \frac{a}{2}\vec{u} \Rightarrow r_1 = r\left(1 + \frac{a^2}{4r^2} + \frac{a}{r}\cos\theta\right)^{1/2}$$

$$\vec{r_2} = \vec{r} - \frac{a}{2}\vec{u} \Rightarrow r_2 = r\left(1 + \frac{a^2}{4r^2} - \frac{a}{r}\cos\theta\right)^{1/2}$$

Using [1.29], we get:

$$V(r,\theta) = \frac{kq}{r}\left\{\left(1 + \frac{a^2}{4r^2} - \frac{a}{r}\cos\theta\right)^{-1/2} - \left(1 + \frac{a^2}{4r^2} + \frac{a}{r}\cos\theta\right)^{-1/2}\right\} \quad [1.30]$$

Since we wish to determine the potential $V(r,\theta)$ and the field $\vec{E}(r,\theta)$ at point M, $(r >> a)$ far from O, an approximation is needed.

The term between parentheses in [1.27] are therefore written approximatively:

$$\begin{cases} \left(1 + \frac{a^2}{4r^2} + \frac{a}{r}\cos\theta\right)^{-\frac{1}{2}} \approx r\left(1 - \frac{a}{2r}\cos\theta\right) \\ \left(1 + \frac{a^2}{4r^2} - \frac{a}{r}\cos\theta\right)^{-\frac{1}{2}} \approx r\left(1 + \frac{a}{2r}\cos\theta\right) \end{cases}$$

Using these approximations, [1.30] becomes:

$$V(r,\theta) = kq\left\{\left(1 + \frac{a}{2r}\cos\theta\right) - \left(1 - \frac{a}{2r}\cos\theta\right)\right\}$$

So, after simplification:

$$V(r,\theta) = \frac{kqa\cos\theta}{r^2} \quad [1.31]$$

Using expression [1.28], the potential [1.31] is expressed as a function of the electric dipole moment. So:

$$V(r,\theta) = \frac{kp\cos\theta}{r^2} \quad [1.32]$$

Now we express the components of the electrostatic field using equation $\vec{E} = -\vec{\nabla} V$ in spherical coordinates. According to [1.9], we find:

$$\begin{cases} E_r = -\dfrac{\partial V}{\partial r} \\ E_\theta = -\dfrac{1}{r}\dfrac{\partial V}{\partial \theta} \end{cases} \Rightarrow \begin{cases} E_r = \dfrac{2kp\cos\theta}{r^3} \\ E_\theta = \dfrac{kp\sin\theta}{r^3} \end{cases} \quad [1.33]$$

Using [1.33], we express the electrostatic field modulus:

$$E = \sqrt{E_r^2 + E_\theta^2} = \frac{kp}{r^3}\sqrt{\left(4\cos^2\theta + \sin^2\theta\right)}$$

So:

$$E = \frac{kp}{r^3}\sqrt{3\cos^2\theta + 1} \quad [1.34]$$

In addition, it would be interesting to express the potential [1.32] and the field [1.34] in Cartesian coordinates. According to Figure 1.7a:

$$r^2 = x^2 + y^2 \; ; \frac{x}{r} = \cos\theta \; ; \sin\theta = \frac{y}{r} \quad [1.35]$$

Using [1.35], the potential [1.32] is written in Cartesian coordinates as:

$$V(x,y) = kp\frac{x}{(x^2+y^2)^{3/2}} \quad [1.36]$$

In Cartesian coordinates, equation $\vec{E} = -\vec{\nabla} V$ gives:

$$E_x = -\frac{\partial V}{\partial x}\ ;\ E_y = -\frac{\partial V}{\partial y} \qquad [1.37]$$

Using [1.37] and expression [1.36], we express the electrostatic field in Cartesian coordinates. We then obtain:

$$\begin{cases} E_x = kp\left\{\dfrac{3x^2}{(x^2+y^2)^{5/2}} - \dfrac{1}{(x^2+y^2)^{3/2}}\right\} \\ E_y = kp\dfrac{3xy}{(x^2+y^2)^{5/2}} \end{cases} \qquad [1.38]$$

Using [1.35], the components [1.38] of the electrostatic field are written as:

$$\begin{cases} E_x = \dfrac{1}{4\pi\varepsilon_0}\dfrac{p(3\cos^2\theta - 1)}{r^3} \\ E_y = \dfrac{1}{4\pi\varepsilon_0}\dfrac{3p\sin\theta\cos\theta}{r^3} \end{cases} \qquad [1.39]$$

1.1.1.7. *Fundamental laws of electrostatics*

The *fundamental laws of electrostatics* are made up of all the laws verified by the electrostatic field. These laws are summarized in Table 1.1.

Circulation of the electrostatic field	**Rotational of the electrostatic field**	**Electrostatic field and potential equation**
$\oint \vec{E} \cdot \vec{dl} = 0$	$\vec{\nabla} \wedge \vec{E} = \vec{0}$	$\vec{E} = -\vec{\nabla} V$
Integral form of Gauss's theorem	**Local form of Gauss's theorem**	**Poisson's law**
$\iint_S \vec{E} \cdot \vec{dS} = \dfrac{Q_{int}}{\varepsilon_0}$	$\vec{\nabla} \cdot \vec{E} = \dfrac{\rho}{\varepsilon_0}$	$\Delta V + \dfrac{\rho}{\varepsilon_0} = 0$

Table 1.1. *Fundamental laws of electrostatics*

In practice, the most appropriate formulation is applied to the problem studied.

1.1.2. *Magnetostatics*[2]

1.1.2.1. *Lorentz's magnetic force, Laplace force*

Let's consider a point charge q in motion with velocity $\vec{v}$ in an area of space where a magnetic field $\vec{B}$ prevails. This charge is subjected to the Lorentz force $\vec{f}$ given by:

$$\vec{f} = q\vec{v} \wedge \vec{B} \qquad [1.40]$$

Lorentz's law [1.40] is defined at the microscopic scale. This allows us to define the magnetic field $\vec{B}$ at the considered point.

Let us now consider a set of point charges in overall motion in a portion of a conductor with length l, placed in a magnetic field $\vec{B}$. We are looking, macroscopically, for the expression of the magnetic force acting on the conductor element l. The overall motion of the charges with velocity $\vec{v}$ corresponds to the circulation of a current of intensity I. During the time period t, the amount of electricity circulating in the portion of conductor of length l is $Q = It$. So $\vec{f_i}$ is the *Lorentz force* acting on a charge q_i. Using [1.40], the result $\vec{F}$ of the magnetic forces acting on element l of the conductor is given by the superposition principle:

$$\vec{F} = \sum_i \vec{f}_i = \sum_i q_i(\vec{v} \wedge \vec{B}) = Q\vec{v} \wedge \vec{B} = I(t\vec{v}) \wedge \vec{B}$$

Knowing that each charge q_i travels a distance $l = vt$, the equation above gives:

$$\vec{F} = I\vec{l} \wedge \vec{B} \qquad [1.41]$$

Expression [1.41], translating the magnetic force acting on a conductor l crossed by a current of intensity I and placed in a magnetic field $\vec{B}$, corresponds to Laplace's law. We obtain this law from the current density vector, in section 1.1.2.2.

2 (Bok and Hulin-Jung 1979; Bruneaux *et al.* 2002; Amazallage *et al.* 2006; Benson 2015; Sakho 2018).

Hendrik Antoon Lorentz was a Dutch physicist. He was renowned for the famous Lorentz spatial transformation laws (transformations of the space-time four-vector, electric and magnetic fields; during a change of reference frame) which are the basis of special relativity. In addition, we owe to him our knowledge of an electromagnetic force acting on a charged particle in motion in an electromagnetic field. Induction due to the movement of a conductor in a time independent magnetic field is called Lorentz induction in his honor.

Box 1.6. *Lorentz (1853–1928)*

1.1.2.2. *Current density vector, continuity equations*

Let us consider any medium (material or vacuum) with volume V along with a volume charge distribution. To establish the continuity equation that translates the charge conservation principle, we add the position $\vec{r}$ of point M in volume V (Figure 1.8) and time t when writing the volume charge density equation $\rho = \rho(\vec{r},t)$. The charge dq contained in the elementary volume dV about point M at time t is $dq = \rho(\vec{r},t)dV$.

The total charge contained in volume V is:

$$q = \iiint_V dq = \iiint_V \rho(\vec{r},t)dV \qquad [1.42]$$

If q depends only on time, the density ρ depends on both the position $\vec{r}$ and time. The differential with respect to time of equation [1.42] gives:

$$\frac{dq}{dt} = \iiint_{(V)} \frac{\partial\rho(\vec{r},t)}{\partial t} dV \qquad [1.43]$$

When the charges are mobile, $\vec{v}$ is the velocity of all charges found around M. By definition, the current density vector often written $\vec{J}(\vec{r},t)$ is given by:

$$\vec{J}(\vec{r},t) = \rho(\vec{r},t)\vec{v} \qquad [1.44]$$

In addition, the motion of the charge carriers causes the circulation of a current of intensity I. By definition, the intensity I is equal to the flux of the current density vector across a surface dS with volume V (Figure 1.8), i.e.:

$$I = \frac{dq}{dt} = \iint_{(S)} \vec{J}(\vec{r},t) \cdot \overrightarrow{dS} = \iint_{(S)} \vec{J}(\vec{r},t) \cdot \vec{n}\, dS \qquad [1.45]$$

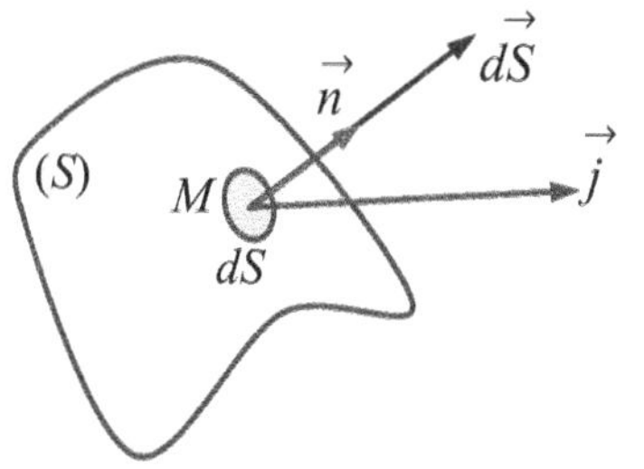

Figure 1.8. *Current density vector at point M*

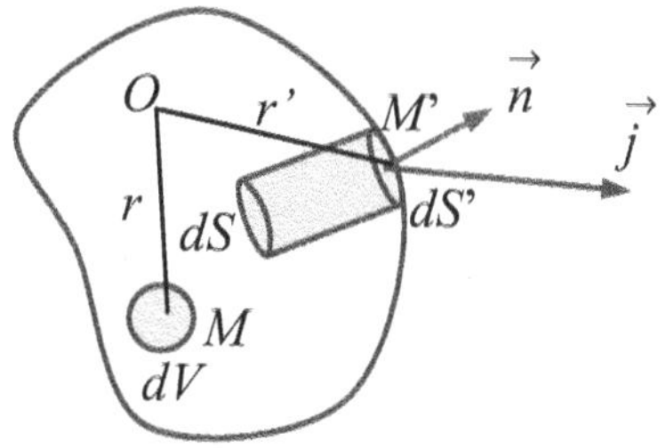

Figure 1.9. *Current density vector at point M'*

The circulation of the current with an intensity I is due to the variation dq of the electric charge contained in volume V during dt. The variation dq is due to the charge dq' leaving the volume V via the elementary surface dS (Figure 1.9) between times t and $t + dt$. The *law of conservation of charge* at time dt needs the following equation to be verified:

$$dq + dq' = 0 \qquad [1.46]$$

With the charge dq having been given in [1.46], we now need to express dq'.

To express dq', we refer to Figure 1.9. The current density vector at point M' is $\vec{J} = \vec{J}(\vec{r}',t)$. The charge dq' leaving the volume V by the base surface dS of cylinder (Figure 1.9) between times t and $t + dt$ is given by an equation that is analogous to [1.46]:

$$\frac{dq'}{dt} = \iint_{(S)} \vec{J}(\vec{r}',t) \cdot \vec{dS} = \iint_{(S)} \vec{J}(\vec{r}',t) \cdot \vec{n}\, dS \qquad [1.47]$$

Using equations [1.45] and [1.47], the *law of conservation of charge* [1.46] is written as:

$$\iint_{(S)} \vec{J}(\vec{r},t) \cdot \vec{dS} + \iint_{(S)} \vec{J}(\vec{r}',t) \cdot \vec{dS} = 0 \qquad [1.48]$$

To account for the total charge q' leaving volume V (Figure 1.9), let us turn into account the *divergence theorem* (*Ostrogradsky's theorem*):

$$\iint_{(S)} \vec{J}(\vec{r}',t) \cdot \vec{dS} = \iiint_{(V)} \vec{\nabla} \cdot \vec{J}(\vec{r},t) dV \qquad [1.49]$$

So, taking into account [1.43] and [1.45], we see that:

$$\iint_{(S)} \vec{J}(\vec{r},t) \cdot \vec{dS} = \iiint_{(V)} \frac{\partial \rho(\vec{r},t)}{\partial t} dV \qquad [1.50]$$

Using [1.49] and [1.50], the law of conservation [1.48] is written in the form:

$$\iiint_{(V)} \vec{\nabla} \cdot \vec{J}(\vec{r},t) dV + \iiint_{(V)} \frac{\partial \rho(\vec{r},t)}{\partial t} dV = 0$$

i.e.:

$$\iiint_{(V)} \left[\vec{\nabla} \cdot \vec{J}(\vec{r},t) + \frac{\partial \rho(\vec{r},t)}{\partial t} \right] dV = 0$$

With this equation being valid whatever the volume V considered, we find the continuity equation reflecting the conservation of electric charge:

$$\vec{\nabla} \cdot \vec{J}(\vec{r},t) + \frac{\partial \rho(\vec{r},t)}{\partial t} = 0 \qquad [1.51]$$

NOTE.– As we pointed out in section 1.1.2.1, Laplace's law [1.41] can be established from the current density vector. For this, let us consider a portion $\overrightarrow{dl}$ of a conductive wire with straight section S traversed by a current with intensity I. This is placed into a magnetic field $\vec{B}$ (Figure 1.10). Here n is the density of electrons per unit volume.

The charge of an electron $q_i = -e$, therefore the volume charge density $\rho = -ne$. The current density vector [1.44] and the intensity I of the current [1.45] are written as:

$$\vec{J} = \rho\vec{v} = -ne\vec{v}; \; I = \iint_{(S)} \vec{J} \cdot \overrightarrow{dS} = -ne\iint_{(S)} \vec{v} \cdot \vec{n}\, dS = -ne\vec{v} \cdot \vec{n} \iint_{(S)} dS$$

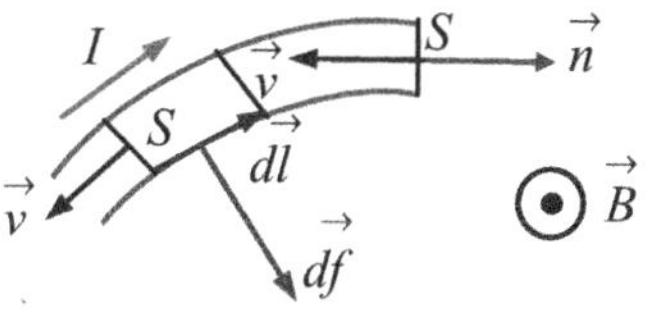

Figure 1.10. *Portion of a conductor with section s, traversed by a current with intensity I*

Knowing that $\vec{n}$ and $\vec{v}$ are anticolinear (Figure 1.10), the intensity of the current is equal to:

$$I = nevS \qquad [1.52]$$

The Lorentz force $\overrightarrow{df_i}$ acting on an electron $q_i = -e$ is:

$$\overrightarrow{df_i} = q_i\vec{v} \wedge \vec{B} = -e\vec{v} \wedge \vec{B} \qquad [1.53]$$

The total number of electrons dN in volume $dV = Sdl$ is $dN = nSdl$. The Lorentz force acting on dN electrons is therefore equal to:

$$\overrightarrow{df} = \sum_i \overrightarrow{df_i} = \sum_i q_i\vec{v} \wedge \vec{B} = -edN\vec{v} \wedge \vec{B} = -neSdl\vec{v} \wedge \vec{B}$$

If $\vec{u}$ is the unit vector according to the direction of the velocity vector $\vec{v}$ ($\vec{u}$ is not represented to not overburden Figure 1.9). Therefore: $\vec{v} = v\vec{u}$ and $\overrightarrow{dl} = -dl\vec{u}$. So, considering [1.52], equation above is written as:

$$\overrightarrow{df} = -neSdl\vec{v} \wedge \vec{B} = nevS\overrightarrow{dl} \wedge \vec{B} = I\overrightarrow{dl} \wedge \vec{B} \quad [1.54]$$

We find the expression of the *Laplace force* acting on the conductor element $\overrightarrow{dl}$ placed in a magnetic field $\vec{B}$.

1.1.2.3. *Biot–Savart law*

The *Biot–Savart law* is used to determine the magnetic field created by a circuit crossed by an electric current of intensity I (Figure 1.11). An element with a length $\overrightarrow{dl}$ of the circuit situated at point M is orientated in the sense of the current circulation. Element $\overrightarrow{dl}$ is small enough to resemble a rectilinear portion. The magnetic field $\overrightarrow{dB}$ created by element $\overrightarrow{dl}$ has a sense such that the trihedron ($\overrightarrow{dl}$, $\vec{u}$, $\overrightarrow{dB}$) is direct. The magnetic field $\overrightarrow{dB}$ created at point P is proportional to I/r^2 and is expressed using equation:

$$\overrightarrow{dB} = \frac{\mu_0}{4\pi} \frac{I\overrightarrow{dl} \wedge \vec{u}}{r^2} \quad [1.55]$$

Considering all the infinitesimal contributions of elements with a length dl_i, we express the magnetic field created by the circuit (C), i.e.:

$$\vec{B} = \frac{\mu_0}{4\pi} \int_{(C)} \frac{I\overrightarrow{dl} \wedge \vec{u}}{r^2} \quad [1.56]$$

The fundamental equation [1.56] describes the *Biot–Savart law*. In the case where there are several independent circuits, vector addition is then extended to all circuits.

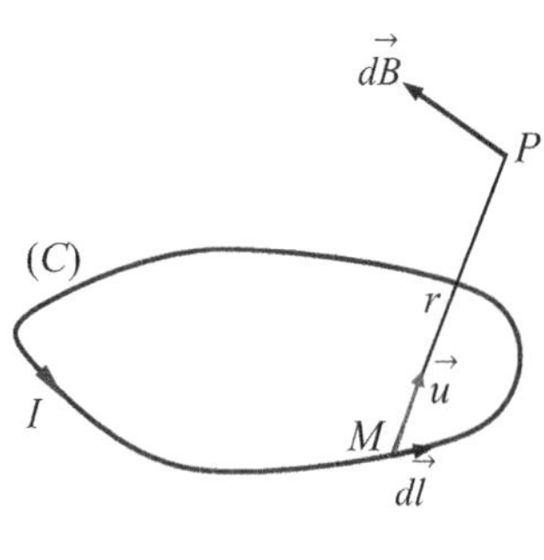

Figure 1.11. *Magnetic field $\overrightarrow{dB}$ created by an element with a length $\overrightarrow{dl}$ of a circuit (C)*

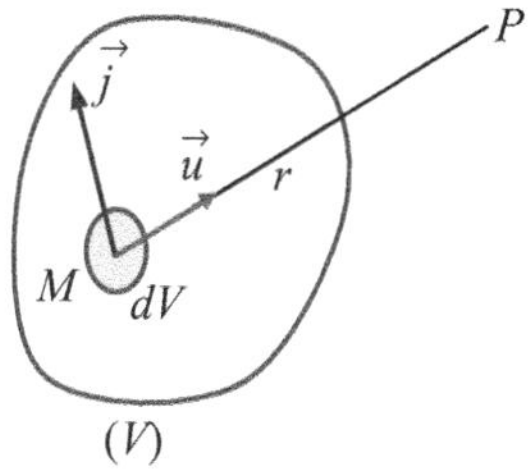

Figure 1.12. *Current distribution within volume V*

Let us consider the case of a current distribution within volume V (Figure 1.12). For this distribution, the field created at point P is given by the *Biot–Savart law* expressed in the form:

$$\vec{B} = \frac{\mu_0}{4\pi} \iiint_{(V)} \frac{\vec{j} \wedge \vec{u}}{r^2} dV \quad [1.57]$$

> **Jean-Baptiste Biot** was a French physicist, astronomer and mathematician. Biot formulated, with his collaborator Félix Savart, the Biot–Savart law, making it possible to determine the magnetic field created at a given point in space, by an electric current passing through a conductor.
>
> **Félix Savart** was a French surgeon and physicist. He became famous in physics for the Biot–Savart law in particular.

Box 1.7. *Biot (1774–1862) and Savart (1791–1841)*

1.1.2.4. *The local law verified by the magnetic field, magnetic field flux*

In the case of the electrostatic field, we have established the local form of Gauss's theorem [1.22] constituting one of Maxwell's fundamental equations in a vacuum. We propose, in what follows, to obtain the magnetostatic equivalent of this theorem. To do this, let us express the divergence of the magnetic field using [1.57]. We obtain:

$$\vec{\nabla} \cdot \vec{B} = \frac{\mu_0}{4\pi} \iiint_{(V)} \vec{\nabla} \cdot \frac{\vec{J} \wedge \vec{u}}{r^2} dV = \frac{\mu_0}{4\pi} \iiint_{(V)} \vec{\nabla} \cdot \left(\vec{J} \wedge \frac{\vec{u}}{r^2} \right) dV$$

Let us consider the vector analysis equation:

$$\vec{\nabla} \cdot (\vec{f} \wedge \vec{g}) = \vec{g} \cdot (\vec{\nabla} \wedge \vec{f}) - \vec{f} \cdot (\vec{\nabla} \wedge \vec{g})$$

The divergence of the magnetic field is written as:

$$\vec{\nabla}\cdot\vec{B}=\frac{\mu_0}{4\pi}\iiint_{(V)}\left[\frac{\vec{u}}{r^2}\cdot(\vec{\nabla}\wedge\vec{J})-\vec{J}\cdot\left(\vec{\nabla}\wedge\frac{\vec{u}}{r^2}\right)\right]dV \quad [1.58]$$

The divergence of a vector $\vec{f}$ in spherical coordinates is given by:

$$\vec{\nabla}\cdot\vec{f}=\frac{1}{r^2}\frac{\partial(r^2 f_r)}{\partial r}+\frac{1}{r\sin\theta}\frac{\partial(\sin\theta\, f_\theta)}{\partial\theta}+\frac{1}{r\sin\theta}\frac{\partial f_\varphi}{\partial\varphi}$$

Let's put $\vec{f}=\frac{\vec{u}}{r}$. Then we obtain:

$$\vec{\nabla}\cdot\vec{f}=\vec{\nabla}\cdot\left(\frac{\vec{u}}{r}\right)=\frac{1}{r^2}\frac{d}{dr}\left(r^2\times\frac{1}{r}\right)=\frac{1}{r^2}\frac{d}{dr}(r)=\frac{1}{r^2}$$

This result allows us to write (the rotational of a gradient being zero):

$$\vec{\nabla}\cdot\left(\frac{\vec{u}}{r}\right)\times\vec{u}=\frac{\vec{u}}{r^2}\Rightarrow\vec{\nabla}\wedge\frac{\vec{u}}{r^2}=\vec{\nabla}\wedge\left[\vec{\nabla}\cdot\frac{\vec{u}}{r}\right]\times\vec{u}=\vec{\nabla}\wedge\vec{\nabla}\left(\frac{1}{r}\right)=\vec{\nabla}\wedge\vec{\nabla}f=\vec{0}$$

In addition, the magnetic field is created at *P*. Therefore the nabla only acts on the coordinates of point *P* and not on those of point *M* or those of the current density vector (Figure 1.12). So, $\vec{\nabla}\wedge\vec{J}=\vec{0}$. With that being said, [1.59] is written as follows:

$$\vec{\nabla}\cdot\vec{B}=0 \quad [1.59]$$

The *local law* [1.59] is general. Whatever the current distribution, the divergence of the magnetic field is equal to zero. Using the divergence theorem, we establish the macroscopic equivalent of local law [1.59]. The flux of the magnetic field through a closed surface *S* enveloping the volume *V* is equal to:

$$\iint_{(S)}\vec{B}\cdot\vec{dS}=\iiint_{(V)}(\vec{\nabla}\cdot\vec{B})dV=0 \quad [1.60]$$

Thus, whatever the current distribution, the flux of the magnetic field through a closed surface *S* is zero. This integral law is general.

1.1.2.5. *Circulation of the magnetic field, Ampère's theorem*

Let us take closed contour (*C*) on which surface (*S*) is based. The current lines are randomly orientated, as shown on Figure 1.13.

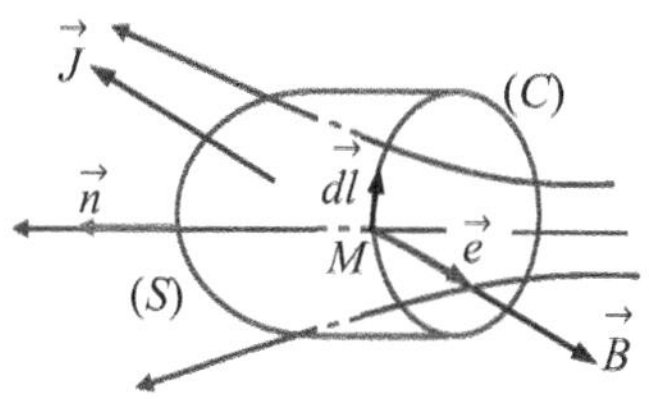

Figure 1.13. *Circulation of a magnetic field on a contour (C)*

The circulation of a magnetic field on a contour (*C*) is equal to:

$$C = \oint_{(C)} \vec{B} \cdot \vec{dl} \qquad [1.61]$$

According to *Ampère's theorem*, the circulation of the magnetic field along a closed contour is equal to the product of the magnetic permeability μ_0 of the vacuum by the algebraic sum I of the current intensities crossing the contour. Therefore:

$$\oint_{(C)} \vec{B} \cdot \vec{dl} = \mu_0 I \qquad [1.62]$$

If $\vec{n}$ is the unit vector normal to the surface based on contour (*C*) and if $\vec{e}$ indicates the sense of the magnetic field, the algebraic sum I in [1.62] is positive if $\vec{n} \cdot \vec{e} > 0$ and negative if $\vec{n} \cdot \vec{e} < 0$.

NOTE.– With the sense of the elementary displacement $\vec{dl}$ being that of the orientation chosen arbitrarily on the contour (*C*), the sense of the unit vector $\vec{n}$ is then defined starting from the direction of the vector $\vec{dl}$ by applying, for example, the right-hand rule.

1.1.2.6. *Rotational of the magnetic field, local formulation of Ampère's law*

In section 1.1.1, we showed that the rotational of the electrostatic field is zero. In this section, we aim to find the magnetostatic equivalent of this rotational. Using [1.62], we obtain the following by applying Stokes' law:

$$\oint_{(C)} \vec{B} \cdot \vec{dl} = \iint_{(S)} (\vec{\nabla} \wedge \vec{B}) \cdot \vec{dS} = \mu_0 I \qquad [1.63]$$

By replacing, in [1.63], the intensity I with its expression [1.45], in which the variables r and t are omitted, we get:

$$\iint_{(S)} (\vec{\nabla} \wedge \vec{B}) \cdot \overrightarrow{dS} = \mu_0 \iint_{(S)} \vec{J} \cdot \overrightarrow{dS}$$

This equality is true whatever the contour (C) and the surface (S) based on it, we ultimately obtain:

$$\vec{\nabla} \wedge \vec{B} = \mu_0 \vec{J} \qquad [1.64]$$

Equation [1.64] expresses the *local form of Ampère's law*. It is one of Maxwell's fundamental equations in steady state processes.

1.1.2.7. *Vector potential, Coulomb gauge*

Knowing that the divergence of a rotational is zero from the local law [1.59], the magnetic field derives from the rotational of a vector field noted $\vec{A}$.

This vector field is known as the vector potential defined by the equation:

$$\vec{B} = \vec{\nabla} \wedge \vec{A} \qquad [1.65]$$

To express the vector potential, we will proceed by analogy with the scalar potential.

Using the local law [1.64], equation [1.65] gives:

$$\vec{\nabla} \wedge \vec{B} = \vec{\nabla} \wedge \vec{\nabla} \wedge \vec{A} = \mu_0 \vec{J} \qquad [1.66]$$

Let us consider the following equation of the vector analysis:

$$\vec{f} \wedge \vec{g} \wedge \vec{h} = \vec{g} \cdot (\vec{f} \cdot \vec{h}) - \vec{h} \cdot (\vec{f} \cdot \vec{g})$$

Equation [1.66] is transformed as follows:

$$\vec{\nabla} \wedge \vec{\nabla} \wedge \vec{A} = \vec{\nabla} \cdot (\vec{\nabla} \cdot \vec{A}) - \vec{A} \cdot (\vec{\nabla} \cdot \vec{\nabla}) = \mu_0 \vec{J}$$

i.e.:

$$\vec{\nabla} \cdot (\vec{\nabla} \cdot \vec{A}) - (\vec{\nabla} \cdot \vec{\nabla}) \cdot \vec{A} = \mu_0 \vec{J}$$

By introducing the Laplacian, we find:

$$\vec{\nabla}\cdot(\vec{\nabla}\cdot\vec{A})-\Delta\vec{A}=\mu_0\vec{J} \qquad [1.67]$$

Equation [1.65] does not allow us to clearly define the vector potential $\vec{A}$. Knowing that the rotational of a gradient is zero, the expression [1.65] is also verified if we change $\vec{A}$ into $\vec{A}_0+\vec{\nabla}\Phi$.

To define a vector potential best adapted to the problem studied (i.e. the vector potential that simplifies the calculation the greatest), we introduce a condition on the vector potential known as the *gauge fixing*.

The vector potential is therefore defined as the vector field satisfying equation [1.65] and the *Coulomb gauge* given by the equation:

$$\vec{\nabla}\cdot\vec{A}=0 \qquad [1.68]$$

Considering the Coulomb gauge, equation [1.67] becomes:

$$\Delta\vec{A}+\mu_0\vec{J}=0 \qquad [1.69]$$

The differential equation [1.69] is analogous to Poisson's equation [1.64]:

$$\Delta V+\frac{\rho}{\varepsilon_0}=0$$

The solution of this equation is formed according to [1.15b] (here, we denote the volume by τ and not by V to avoid confusion with the potential V):

$$V=\iiint_{(\tau)}\frac{\rho\,d\tau}{4\pi\varepsilon_0\,r} \qquad [1.70a]$$

The solution of the differential equation [1.70] is then in the following form:

$$\vec{A}=\iiint_{(\tau)}\frac{\mu_0\vec{J}\,d\tau}{4\pi\,r} \qquad [1.70b]$$

In [1.70b], $\vec{A}$ designates the vector potential created at point P in space by the distribution of current $\vec{J}$ inside volume τ (Figure 1.12).

We can determine [1.70b] from a direct calculation using the Biot–Savart law [1.67]. For this, we rewrite the magnetic field in the following form:

$$\vec{B} = \frac{\mu_0}{4\pi}\iiint_{(\tau)} \frac{\vec{J} \wedge \vec{u}}{r^2} dV = -\frac{\mu_0}{4\pi}\iiint_{(\tau)} \frac{\vec{u}}{r^2} \wedge \vec{J}\, d\tau \qquad [1.71]$$

The last equality in equation [1.71] is justified by the property of the vector product $\vec{f} \wedge \vec{g} = -\vec{g} \wedge \vec{f}$. Expression [1.70b] shows that the vector potential is the integral of a function in $1/r$. This transforms the term in $1/r^2$ in [1.71] into a function in $1/r$. For this, we express the gradient of the scalar function $f(r) = 1/r$. In spherical coordinates, we get:

$$\vec{\nabla} f = \frac{\partial f}{\partial r}\vec{e}_r + \frac{1}{r}\frac{\partial f}{\partial \theta}\vec{e}_\theta + \frac{1}{r\sin\theta}\frac{\partial f}{\partial \varphi}\vec{e}_\varphi$$

Using this equation, we find:

$$\vec{\nabla} f = \vec{\nabla}\left(\frac{1}{r}\right) = \frac{d}{dr}\left(\frac{1}{r}\right)\vec{e}_r = -\frac{1}{r^2}\vec{e}_r$$

Knowing that in [1.71] $\vec{e}_r = \vec{u}$, we can write:

$$-\frac{\vec{u}}{r^2} \wedge \vec{J} = \vec{\nabla}\left(\frac{1}{r}\right) \wedge \vec{J}$$

Expression [1.71] then gives:

$$\vec{B} = \frac{\mu_0}{4\pi}\iiint_{(\tau)} \vec{\nabla}\left(\frac{1}{r}\right) \wedge \vec{J}\, d\tau = \frac{\mu_0}{4\pi}\iiint_{(\tau)} \vec{\nabla} f \wedge \vec{j}\, d\tau \qquad [1.72]$$

Let's take into account the vector equation:

$$\vec{\nabla} \wedge (f\vec{g}) = f\vec{\nabla} \wedge \vec{g} + \vec{\nabla} f \wedge \vec{g} \Rightarrow \vec{\nabla} f \wedge \vec{g} = \vec{\nabla} \wedge (f\vec{g}) - f\vec{\nabla} \wedge \vec{g}$$

Considering the last member of [1.72] and the last vector equation above, we can write (bearing in mind that the nabla operator does not act on the current density vector):

$$\vec{\nabla} f \wedge \vec{J} = \vec{\nabla} \wedge (f\vec{J}) - f\vec{\nabla} \wedge \vec{J} = \vec{\nabla} \wedge (f\vec{J}) \Rightarrow \vec{\nabla}\left(\frac{1}{r}\right) \wedge \vec{J} = \vec{\nabla} \wedge \left(\frac{\vec{J}}{r}\right)$$

Using the last equality above, expression [1.72] gives:

$$\vec{B} = \frac{\mu_0}{4\pi} \iiint_{(\tau)} \vec{\nabla} \wedge \frac{\vec{J}}{r} d\tau = \vec{\nabla} \wedge \frac{\mu_0}{4\pi} \iiint_{(\tau)} \frac{\vec{J}}{r} d\tau \qquad [1.73]$$

If we compare [1.73] and [1.65], we find that:

$$\vec{A} = \frac{\mu_0}{4\pi} \iiint_{(\tau)} \frac{\vec{J}}{r} d\tau = \frac{\mu_0}{4\pi} \iiint_{(\tau)} \frac{I\vec{dl}}{r} d\tau \qquad [1.74]$$

The last equality in equation [1.74] can be established using the Biot–Savart law [1.56].

NOTE.– When studying the properties of electromagnetic waves, instead of the Coulomb gauge one uses the *Lorentz gauge*, which links the vector potential $\vec{A}$ and the scalar potential V (see section 1.1.4.5):

$$\begin{cases} \vec{\nabla} \cdot \vec{A} + \dfrac{1}{c^2} \dfrac{\partial V}{\partial t} = 0 \\ \vec{A}(\infty) = \vec{0} \; ; V(\infty) = 0. \end{cases} \qquad [1.75]$$

1.1.2.8. *Circulation of the vector potential*

The expression of the circulation of the vector potential along a closed contour allow us to determine its physical sense and its unit. Using Stokes' theorem, we get:

$$\oint_{(C)} \vec{A} \cdot \vec{dl} = \iint_{(S)} (\vec{\nabla} \wedge \vec{A}) \cdot \vec{dS} \qquad [1.76]$$

Using equation [1.65] defining the vector potential, [1.76] becomes:

$$\oint_{(C)} \vec{A} \cdot \vec{dl} = \iint_{(S)} \vec{B} \cdot \vec{dS} = \Phi \qquad [1.77]$$

therefore, the circulation of the vector potential along a closed contour (C) is equal to the flux of the magnetic field across a surface (S) based on this contour. Considering equation [1.77], we can write that $Adl = \Phi$. Knowing that the magnetic flux is expressed in weber (Wb), the vector potential is expressed in weber per meter ($\mathrm{Wb} \cdot \mathrm{m}^{-1}$).

1.1.2.9. *Electrostatic dipole/magnetic dipole analogy*

The electrostatic dipole was already studied in section 1.1.1 (Figure 1.7a). To make the electrostatic dipole-magnetic dipole analogy, it is necessary to study the magnetic dipole. Here we consider a circular coil with radius a, axis Oy and traversed by a constant current with intensity I (Figure 1.14).

The coil traversed by a current behaves as a magnetic dipole. The magnetic moment $\vec{M}$ of this dipole is a vector normal to the plane of the coil. Its orientation depends on the positive sense defined by the sense of the traveling current. In what follows, we are also interested in the calculation of the magnetic field at a point P far from the coil ($OP = r >> a$). The chosen plane zOx is that of the coil.

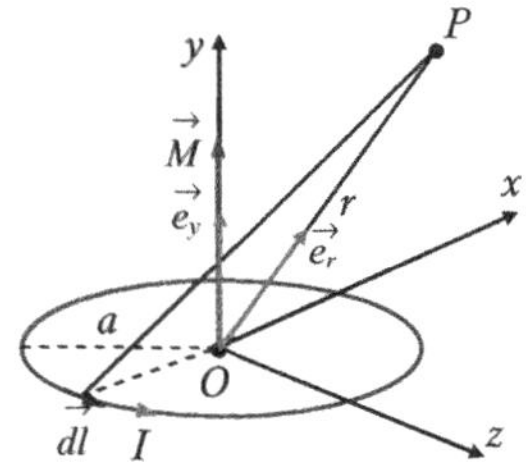

Figure 1.14. *Magnetic dipole composed of a circular coil with magnetic moment M*

By definition, the moment magnetic is given by the equation:

$$\vec{M} = I\vec{S} = \pi a^2 I \vec{e}_y \qquad [1.78]$$

Moment $\vec{M}$ is therefore orientated according to axis Oy. To make the electric dipole-magnetic dipole analogy, below we give the schemas of an electric dipole (Figure 1.15a) and a magnetic dipole (Figure 1.15b) composed of the coil, as shown in Figure 1.14. By definition, the electric dipole moment is given by equation [1.28] where we replaced $\vec{p}$ by $\vec{M}$:

$$\vec{M} = q\overrightarrow{AB} = qa\vec{u} \qquad [1.79]$$

The scalar potential [1.71] can be written in the following form:

$$V = \frac{1}{4\pi\varepsilon_0}\frac{\vec{M}\cdot\vec{e}_r}{r^2} \qquad [1.80]$$

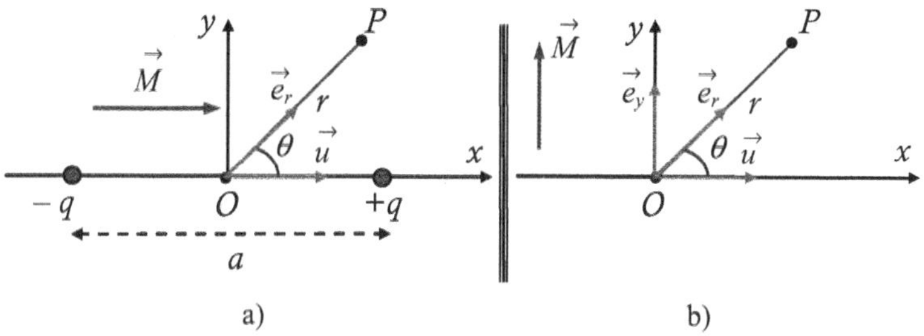

Figure 1.15. *Analogy a) electric dipole; b) magnetic dipole*

By making the analogy $1/4\pi\varepsilon_0 \leftrightarrow \mu_0/4\pi$, the vector potential can be written in the following form [1.80]:

$$\vec{A} = \frac{\mu_0}{4\pi}\frac{\vec{M} \wedge \vec{e}_r}{r^2} \qquad [1.81]$$

The potential and the components of the electrostatic field created by the electrostatic dipole (Figure 1.15a) in cylindrical coordinates are given by expressions [1.32] and [1.33], in which we replace p with qa:

$$V = \frac{1}{4\pi\varepsilon_0}\frac{qa\cos\theta}{r^2}$$

$$E_r = \frac{1}{4\pi\varepsilon_0}\frac{qa2\cos\theta}{r^3}; E_\theta = \frac{1}{4\pi\varepsilon_0}\frac{qa\sin\theta}{r^3}; E_z = 0 \qquad [1.82]$$

What is left is to determine the components of the vector potential and the magnetic field vector. According to Figure 1.15b, knowing that $\vec{u} = \vec{e}_x$, we obtain:

$$\vec{M} \wedge \vec{e}_r = M\vec{e}_y \wedge (\cos\theta\vec{e}_x + \sin\theta\vec{e}_y) = M\cos\theta\vec{e}_y \wedge \vec{e}_x$$

$$= -M\cos\theta\vec{e}_z = -I\pi a^2\cos\theta\vec{e}_z$$

The components of the vector potential [1.81] are therefore equal to:

$$A_r = 0 ; A_\theta = 0 ; A_z = -\frac{\mu_0}{4\pi}\frac{I\pi a^2\cos\theta}{r^2} \qquad [1.83]$$

The magnetic field is linked to the vector potential by the equation $\vec{B} = \vec{\nabla} \wedge \vec{A}$. Moreover, the components of the vector field rotational $\vec{f}$ in cylindrical coordinates are written:

$$\vec{\nabla} \wedge \vec{f} = \left(\frac{1}{r} \frac{\partial f_z}{\partial \theta} - \frac{\partial f_\theta}{\partial z} \right) \vec{e}_r + \left(\frac{\partial f_r}{\partial z} - \frac{\partial f_z}{\partial r} \right) \vec{e}_\theta + \frac{1}{r} \left(\frac{\partial (r f_\theta)}{\partial r} - \frac{\partial f_r}{\partial \theta} \right) \vec{e}_z$$

Using this equation and the components [1.83] of the vector potential, we determine the components of the magnetic field as follows:

$$\begin{cases} B_r = \left(\dfrac{1}{r} \dfrac{\partial A_z}{\partial \theta} - \dfrac{\partial A_\theta}{\partial z} \right) = -\dfrac{\partial A_\theta}{\partial z} = \dfrac{\mu_0}{4\pi} \dfrac{I\pi a^2 \sin\theta}{r^3} \\ B_\theta = \left(\dfrac{\partial A_r}{\partial z} - \dfrac{\partial A_z}{\partial r} \right) = -\dfrac{\partial A_z}{\partial r} = \dfrac{\mu_0}{4\pi} \dfrac{I\pi a^2 2\cos\theta}{r^3} \end{cases} ; B_z = \frac{1}{r} \left(\frac{\partial (r A_\theta)}{\partial r} - \frac{\partial A_r}{\partial \theta} \right) = 0 \quad [1.84]$$

The electrostatic dipole/magnetic dipole analogy is presented in Table 1.2.

	Electric variable		Magnetic variable
Electric dipole moment	$M = qa$	Magnetic dipole moment	$M = I\pi a^2$
Electric potential	$V = \dfrac{1}{4\pi\varepsilon_0} \dfrac{\overrightarrow{M} \cdot \overrightarrow{e_r}}{r^2}$	Vector potential	$\vec{A} = \dfrac{\mu_0}{4\pi} \dfrac{\overrightarrow{M} \wedge \overrightarrow{e_r}}{r^2}$
Electric field	$E_r = \dfrac{1}{4\pi\varepsilon_0} \dfrac{qa\,2\cos\theta}{r^3}$	Magnetic field	$B_r = \dfrac{\mu_0}{4\pi} \dfrac{I\pi a^2 \sin\theta}{r^3}$
	$E_\theta = \dfrac{1}{4\pi\varepsilon_0} \dfrac{qa \sin\theta}{r^3}$		$B_\theta = \dfrac{\mu_0}{4\pi} \dfrac{I\pi a^2 2\cos\theta}{r^3}$
	$E_z = 0$		$B_z = 0$

Table 1.2. *Electrostatic dipole/magnetic dipole*

1.1.2.10. *Fundamental laws of magnetostatics*

The fundamental laws of magnetostatics established in this section are summarized in Table 1.3.

Circulation of the magnetic field (integral form of Ampère's law)	Rotational of the magnetic field (local form of Ampère's law)	Magnetic field and vector potential equation
$\int_{(C)} \vec{B} \cdot \vec{dl} = \mu_0 I$	$\vec{\nabla} \wedge \vec{B} = \mu_0 \vec{j}$	$\vec{B} = \vec{\nabla} \wedge \vec{A}$
Magnetic field flux	**Local law of magnetic field**	**Local law of the vector potential**
$\iint_{(S)} \vec{B} \cdot \vec{dS} = 0$	$\vec{\nabla} \cdot \vec{B} = 0$	$\Delta\vec{A} + \mu_0 \vec{j} = \vec{0}$

Table 1.3. *Fundamental laws of magnetostatics*

This table contains several formulations of laws in magnetostatics. In practice, one applies the formulation best suited to the problem studied.

1.1.3. *Electromagnetic induction*[3]

1.1.3.1. *Experimental evidence*

In 1831, Faraday discovered the induction phenomenon from the relative motion between a U-shaped magnet and a coil. In practice, we use a straight magnet, a coil and a galvanometer (Figure 1.16) to highlight this phenomenon.

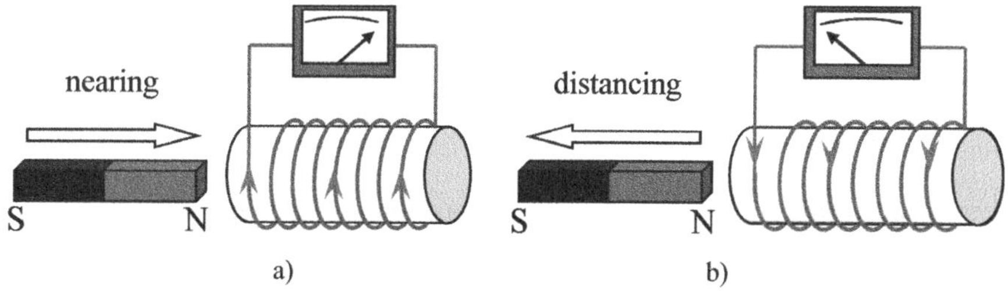

Figure 1.16. *a) Magnet far from the coil axis: an electric current occurs; b) magnet close to the axis of the current: an electric current occurs*

3 (Annequin and Boutigny 1974; Bok and Hulin-Jung 1979; Bruneaux *et al.* 2002; Krempf 2004; Benson 2015; Sakho 2018).

The straight magnet is moved along the axis of a coil (Figure 1.16b) or away from it (Figure 1.16a). According to the direction in which the magnet is moved, the galvanometer needle deviates either to the right (Figure 1.16b) or to the left (Figure 1.16a). This justifies the generation of an electric current of intensity i, whose sense depends on the direction in which the magnet is moving. This current, induced by the displacement of the magnet, is called the induced current. The phenomenon that gives rise to the induced current is called electromagnetic induction. The moving magnet is the inductor and the coil is the armature. By fixing the magnet (the armature) and by moving the coil (the inductor), we observe the same phenomenon: the phenomenon of induction appears during a relative displacement of an inductor and an induced current (Figure 1.16b).

By definition, electromagnetic induction is the phenomenon of producing electrical effects from magnetic fields. As we pointed out earlier, the induced current i changes direction depending on whether one brings the magnet closer or moves it away. The direction of the current i is governed by *Lenz's law*, which states the following: *the direction of the induced current is such that it opposes the change that created it.* When the magnet is brought closer, *the lines of the magnetic field become more and more concentrated across the coil*; the induced current opposes this increase in magnetic flux (Figure 1.16b). When moving the magnet away: *the lines of the magnetic fields decrease across the coil*; the induced current opposes this decrease in flux (Figure 1.16a). The direction of the induced current is therefore linked to the variation in magnetic flux through the coil.

Michael Faraday was a British physicist and chemist. In 1831, Faraday discovered the very important phenomenon of induction from the relative movement between a U-shaped magnet and a coil. He also became famous through the formulation of Faraday's law of induction in 1834 on electromagnetic induction (this law, based on Faraday's work in 1831, was published by Lenz in 1834).

Heinrich Friedrich Emil Lenz was a German physicist. Appointed as professor at the Saint Petersburg Imperial University, there he repeated Faraday's experiments. This allowed him to state the law indicating the direction of the induced current (Lenz's law) in 1833. Lenz is especially famous for this important law relating to electromagnetic induction.

Box 1.8. *Faraday (1791–1867) and Lenz (1804–1865)*

1.1.3.2. *Lagrangian of a particle in an electromagnetic field*

By definition, the electromagnetic field is the sum of the vector fields $(\vec{E},\vec{B})$. The properties of an electromagnetic field at a point in space are determined by the properties of the electric field $\vec{E}$ and the magnetic field $\vec{B}$ at the point considered.

In electromagnetism, it is stated that any charged particle is characterized by its mass m and its electric charge q. These two characteristics are independent of the chosen Galilean reference frame. In addition, for a relativistic particle in motion with velocity $\vec{v}$ in the electromagnetic field, it is defined a physical variable known as the *Lagrangian L*.

It is assumed that:

$$L=-mc^2\sqrt{1-\frac{v^2}{c^2}}+q(\vec{A}\cdot\vec{v}-V) \qquad [1.85]$$

In the second term of equation [1.85]:

– the first term corresponds to the Lagrangian of the free particle (not subject to a force field);

– the second term is the Lagrangian of the particle subject to an electromagnetic field with $\vec{A}$ the four-potential of components: $\vec{A}(A_x,A_y,A_z,V/c)$, $A_x(x, y, z, t)$, A_y $x, y, z, t)$ and A_z $x, y, z, t)$ and $V = V(x, y, z, t)$ is the scalar potential.

1.1.3.3. *Definitions of electric field and magnetic field*

Any point charge q in motion with velocity $\vec{v}$ in the electromagnetic field $(\vec{E},\vec{B})$, is subjected to the electromagnetic force known as the *Lorentz force* defined by the equation:

$$\vec{f}=q(\vec{E}+\vec{v}\wedge\vec{B}) \qquad [1.86]$$

To define the electric field and the magnetic field, one must determine the expression of the electromagnetic force [1.86] from the fundamental equation of dynamics [1.87] and Lagrange equations of the particle [1.88]:

$$\vec{f} = \frac{d\vec{p}}{dt}\ ;\ \vec{p} = \frac{m\vec{v}}{\sqrt{1 - v^2 / c^2}} \qquad [1.87]$$

$$\frac{d}{dt}\frac{\partial L}{\partial u'} - \frac{\partial L}{\partial u} = 0 \qquad [1.88]$$

Equation [1.88] corresponds to the *Lagrange equation* of the particle relative to the generalized coordinate u ($u = x$, y or z); $u' = \frac{du}{dt} = v_u$ where v_u is the velocity component according to u: v_u (v_x, v_y, v_z). Using [1.85], it becomes:

$$L = -mc^2\sqrt{1 - \frac{(v_x + v_y + v_z)^2}{c^2}} + q(A_x v_x + A_y v_y + A_z v_z - V) \qquad [1.89]$$

With the axes Ox, Oy and Oz being equivalent (since space is isotropic), we express the Lagrange equation of the particle relative to coordinate x. Using [1.89], we obtain:

$$\frac{\partial L}{\partial u'} = \frac{\partial L}{\partial v_x} = -mc^2\frac{1}{2}\left(-\frac{2v_x}{c^2}\right) \times \left[1 - \frac{(v_x + v_y + v_z)^2}{c^2}\right]^{-1/2} + qA_x = \frac{m}{\sqrt{1 - \frac{v^2}{c^2}}}\frac{dx}{dt} + qA_x$$

In addition:

$$\frac{\partial L}{\partial u} = \frac{\partial L}{\partial x} = q\left(\frac{\partial A_x}{\partial x}v_x + \frac{\partial A_y}{\partial x}v_y + \frac{\partial A_Z}{\partial x}v_z - \frac{\partial V}{\partial x}\right) = q\left(v_x\frac{\partial A_x}{\partial x} + v_y\frac{\partial A_y}{\partial x} + v_z\frac{\partial A_Z}{\partial x} - \frac{\partial V}{\partial x}\right)$$

Using these two results, the Lagrange equation of the particle relative to coordinate x is written:

$$\frac{d}{dt}\left(\frac{m}{\sqrt{1 - \frac{v^2}{c^2}}}\frac{dx}{dt} + qA_x\right) - q\left(\frac{dx}{dt}\frac{\partial A_x}{\partial x} + \frac{dy}{dt}\frac{\partial A_y}{\partial x} + \frac{dz}{dt}\frac{\partial A_z}{\partial x} - \frac{\partial V}{\partial x}\right) = 0$$

Simplifying this equation, we find:

$$\frac{d}{dt}\left(\frac{mx'}{\sqrt{1-\frac{v^2}{c^2}}}+qA_x\right)-q\left(x'\frac{\partial A_x}{\partial x}+y'\frac{\partial A_y}{\partial x}+z'\frac{\partial A_z}{\partial x}-\frac{\partial V}{\partial x}\right)=0 \qquad [1.90]$$

In the first term, between the parentheses of the right term in equation [1.90], we find the expression of impulsion [1.87] relative to coordinate x:

$$p_x=\frac{mv_x}{\sqrt{1-\frac{v^2}{c^2}}}=\frac{m}{\sqrt{1-\frac{v^2}{c^2}}}\frac{dx}{dt}=\frac{mx'}{\sqrt{1-\frac{v^2}{c^2}}}$$

Now [1.90] is written as:

$$\frac{d}{dt}(p_x+qA_x)-q\left(x'\frac{\partial A_x}{\partial x}+y'\frac{\partial A_y}{\partial x}+z'\frac{\partial A_z}{\partial x}-\frac{\partial V}{\partial x}\right)=0$$

i.e.:

$$\frac{dp_x}{dt}+q\frac{dA_x}{dt}-q\left(x'\frac{\partial A_x}{\partial x}+y'\frac{\partial A_y}{\partial x}+z'\frac{\partial A_z}{\partial x}-\frac{\partial V}{\partial x}\right)=0 \qquad [1.91]$$

Knowing that $A_x = A_x(x, y, z, t)$, the differential of A_x with respect to time is equal to:

$$\frac{dA_x}{dt}=\frac{\partial A_x}{\partial x}\frac{\partial x}{\partial t}+\frac{\partial A_x}{\partial y}\frac{\partial y}{\partial t}+\frac{\partial A_x}{\partial z}\frac{\partial z}{\partial t}+\frac{\partial A_x}{\partial t}=\frac{\partial A_x}{\partial x}x'+\frac{\partial A_x}{\partial y}y'+\frac{\partial A_x}{\partial z}z'+\frac{\partial A_x}{\partial t}$$

Using this result, expression [1.91] gives:

$$\frac{dp_x}{dt}+q\left(\frac{\partial A_x}{\partial x}x'+\frac{\partial A_x}{\partial y}y'+\frac{\partial A_x}{\partial z}z'+\frac{\partial A_x}{\partial t}\right)-q\left(x'\frac{\partial A_x}{\partial x}+y'\frac{\partial A_y}{\partial x}+z'\frac{\partial A_z}{\partial x}-\frac{\partial V}{\partial x}\right)=0$$

Rearranging this equation, we get:

$$\frac{dp_x}{dt}+q\left(\frac{\partial A_x}{\partial x}x'-x'\frac{\partial A_x}{\partial x}+\frac{\partial A_x}{\partial y}y'-y'\frac{\partial A_y}{\partial x}+\frac{\partial A_x}{\partial z}z'-z'\frac{\partial A_z}{\partial x}+\frac{\partial A_x}{\partial t}-\frac{\partial V}{\partial x}\right)=0$$

i.e.:

$$\frac{dp_x}{dt}+qy'\left(\frac{\partial A_x}{\partial y}-\frac{\partial A_y}{\partial x}\right)+qz'\left(\frac{\partial A_x}{\partial z}-\frac{\partial A_z}{\partial x}\right)+q\left(\frac{\partial A_x}{\partial t}+\frac{\partial V}{\partial x}\right)=0$$

Therefore:

$$\frac{dp_x}{dt}=q\left\{-\frac{\partial A_x}{\partial t}-\frac{\partial V}{\partial x}+y'\left(\frac{\partial A_y}{\partial x}-\frac{\partial A_x}{\partial y}\right)-z'\left(\frac{\partial A_x}{\partial z}-\frac{\partial A_z}{\partial x}\right)\right\} \quad [1.92a]$$

By circular permutation, the analogous equations relative to variables y and z are written as follows:

$$\frac{dp_y}{dt}=q\left\{-\frac{\partial A_y}{\partial t}-\frac{\partial V}{\partial y}+z'\left(\frac{\partial A_z}{\partial y}-\frac{\partial A_y}{\partial z}\right)-x'\left(\frac{\partial A_y}{\partial x}-\frac{\partial A_x}{\partial y}\right)\right\} \quad [1.92b]$$

$$\frac{dp_z}{dt}=q\left\{-\frac{\partial A_z}{\partial t}-\frac{\partial V}{\partial z}+x'\left(\frac{\partial A_x}{\partial z}-\frac{\partial A_z}{\partial x}\right)-y'\left(\frac{\partial A_z}{\partial y}-\frac{\partial A_y}{\partial z}\right)\right\} \quad [1.92c]$$

By combining equations [1.92], we get:

$$\frac{d\vec{p}}{dt}=\frac{dp_x}{dt}\vec{e}_x+\frac{dp_y}{dt}\vec{e}_y+\frac{dp_z}{dt}\vec{e}_z \quad [1.93a]$$

$$\frac{\partial\vec{A}}{\partial t}=\frac{\partial A_x}{\partial t}\vec{e}_x+\frac{\partial A_y}{\partial t}\vec{e}_y+\frac{\partial A_z}{\partial t}\vec{e}_z \quad [1.93b]$$

$$\vec{\nabla}V=-\frac{\partial V}{\partial x}-\frac{\partial V}{\partial y}-\frac{\partial V}{\partial z} \quad [1.93c]$$

The rotational of the vector potential is given by:

$$\vec{\nabla} \wedge \vec{A} = \left(\frac{\partial A_z}{\partial y} - \frac{\partial A_y}{\partial z} \right) \vec{e_x} + \left(\frac{\partial A_x}{\partial z} - \frac{\partial A_z}{\partial x} \right) \vec{e_y} + \left(\frac{\partial A_x}{\partial x} - \frac{\partial A_z}{\partial y} \right) \vec{e_z} \qquad [1.94]$$

We will now express the vector double product $\vec{v} \wedge \vec{\nabla} \wedge \vec{A}$ using expression [1.94] for the rotational of the vector potential. We obtain:

$$\vec{v} \wedge \vec{\nabla} \wedge \vec{A} = \vec{v} \wedge \left(\frac{\partial A_z}{\partial y} - \frac{\partial A_y}{\partial z} \right) \vec{e_x} + \vec{v} \wedge \left(\frac{\partial A_x}{\partial z} - \frac{\partial A_z}{\partial x} \right) \vec{e_y} + \vec{v} \wedge \left(\frac{\partial A_x}{\partial x} - \frac{\partial A_z}{\partial y} \right) \vec{e_z} \qquad [1.95]$$

As an example, we express the first expression of the right term of this equation:

$$\vec{v} \wedge \left(\frac{\partial A_z}{\partial y} - \frac{\partial A_y}{\partial z} \right) \vec{e_x} = \left(v_x \vec{e_x} + v_y \vec{e_y} + v_z \vec{e_z} \right) \wedge \left(\frac{\partial A_z}{\partial y} - \frac{\partial A_y}{\partial z} \right) \vec{e_x}$$

i.e.:

$$\vec{v} \wedge \left(\frac{\partial A_z}{\partial y} - \frac{\partial A_y}{\partial z} \right) \vec{e_x} = v_y \left(\frac{\partial A_z}{\partial y} - \frac{\partial A_y}{\partial z} \right) \vec{e_y} \wedge \vec{e_x} + v_z \left(\frac{\partial A_z}{\partial y} - \frac{\partial A_y}{\partial z} \right) \vec{e_z} \wedge \vec{e_x}$$

Therefore:

$$\vec{v} \wedge \left(\frac{\partial A_z}{\partial y} - \frac{\partial A_y}{\partial z} \right) \vec{e_x} = -v_y \left(\frac{\partial A_z}{\partial y} - \frac{\partial A_y}{\partial z} \right) \vec{e_z} + v_z \left(\frac{\partial A_z}{\partial y} - \frac{\partial A_y}{\partial z} \right) \vec{e_y}$$

Introducing the notation $v_u = du/dt = u'$, we obtain:

$$\vec{v} \wedge \left(\frac{\partial A_z}{\partial y} - \frac{\partial A_y}{\partial z} \right) \vec{e_x} = -y' \left(\frac{\partial A_z}{\partial y} - \frac{\partial A_y}{\partial z} \right) \vec{e_z} + z' \left(\frac{\partial A_z}{\partial y} - \frac{\partial A_y}{\partial z} \right) \vec{e_y} \qquad [1.96a]$$

By adopting the same approach, we find, for the other two components of the vector double product [1.95]:

$$\begin{cases} \vec{v} \wedge \left(\frac{\partial A_x}{\partial z} - \frac{\partial A_z}{\partial x} \right) \vec{e_y} = x' \left(\frac{\partial A_x}{\partial z} - \frac{\partial A_z}{\partial x} \right) \vec{e_z} - z' \left(\frac{\partial A_x}{\partial z} - \frac{\partial A_z}{\partial x} \right) \vec{e_x} \\ \vec{v} \wedge \left(\frac{\partial A_x}{\partial x} - \frac{\partial A_z}{\partial y} \right) \vec{e_z} = -x' \left(\frac{\partial A_y}{\partial x} - \frac{\partial A_x}{\partial y} \right) \vec{e_y} + y' \left(\frac{\partial A_y}{\partial x} - \frac{\partial A_x}{\partial y} \right) \vec{e_x} \end{cases} \qquad [1.96b]$$

Using [1.96], equation [1.94] is written as follows:

$$\vec{v}\wedge\vec{\nabla}\wedge\vec{A}=-y'\left(\frac{\partial A_z}{\partial y}-\frac{\partial A_y}{\partial z}\right)\vec{e}_z+z'\left(\frac{\partial A_z}{\partial y}-\frac{\partial A_y}{\partial z}\right)\vec{e}_y+x'\left(\frac{\partial A_x}{\partial z}-\frac{\partial A_z}{\partial x}\right)\vec{e}_z$$
$$-z'\left(\frac{\partial A_x}{\partial z}-\frac{\partial A_z}{\partial x}\right)\vec{e}_x-x'\left(\frac{\partial A_y}{\partial x}-\frac{\partial A_x}{\partial y}\right)\vec{e}_y+y'\left(\frac{\partial A_y}{\partial x}-\frac{\partial A_x}{\partial y}\right)\vec{e}_x$$

We now need to arrange this expression to reveal that of the rotational vector potential. We intuitively obtain:

$$\vec{v}\wedge\vec{\nabla}\wedge\vec{A}=\left[y'\left(\frac{\partial A_y}{\partial x}-\frac{\partial A_x}{\partial y}\right)-z'\left(\frac{\partial A_x}{\partial z}-\frac{\partial A_z}{\partial x}\right)\right]\vec{e}_x+$$
$$\left[z'\left(\frac{\partial A_z}{\partial y}-\frac{\partial A_y}{\partial z}\right)-x'\left(\frac{\partial A_y}{\partial x}-\frac{\partial A_x}{\partial y}\right)\right]\vec{e}_y+\left[x'\left(\frac{\partial A_x}{\partial z}-\frac{\partial A_z}{\partial x}\right)-y'\left(\frac{\partial A_z}{\partial y}-\frac{\partial A_y}{\partial z}\right)\right]\vec{e}_z \quad [1.97]$$

Considering [1.97], the expressions [1.92a] are written, respectively, as follows:

$$\begin{cases}\dfrac{dp_x}{dt}=q\left\{-\dfrac{\partial A_x}{\partial t}-\dfrac{\partial V}{\partial x}+(\vec{v}\wedge\vec{\nabla}\wedge\vec{A})\cdot\vec{e}_x\right\}\\ \dfrac{dp_y}{dt}=q\left\{-\dfrac{\partial A_y}{\partial t}-\dfrac{\partial V}{\partial y}+(\vec{v}\wedge\vec{\nabla}\wedge\vec{A})\cdot\vec{e}_y\right\}\\ \dfrac{dp_z}{dt}=q\left\{-\dfrac{\partial A_z}{\partial t}-\dfrac{\partial V}{\partial z}+(\vec{v}\wedge\vec{\nabla}\wedge\vec{A})\cdot\vec{e}_z\right\}\end{cases} \quad [1.98]$$

Combining [1.93] and [1.98], the *fundamental equation of dynamics* [1.87] is ultimately written. Then we obtain:

$$\vec{f}=\frac{d\vec{p}}{dt}=q\left\{-\frac{\partial\vec{A}}{\partial t}-\vec{\nabla}V+\vec{v}\wedge(\vec{\nabla}\wedge\vec{A})\right\} \quad [1.99]$$

Comparing [1.99] to the *electromagnetic force* [1.86], we deduce the definitions of the electric field and magnetic field.

So, by definition, the electric field and magnetic field at point $\vec{r}$ in space at time t are given by the following expressions:

$$\begin{cases} \vec{E}(\vec{r},t) = -\dfrac{\partial \vec{A}(\vec{r},t)}{\partial t} - \vec{\nabla} V(\vec{r},t) \\ \vec{B}(\vec{r},t) = \vec{\nabla} \wedge \vec{A}(\vec{r},t) \end{cases} \Rightarrow \begin{cases} \vec{E} = -\dfrac{\partial \vec{A}}{\partial t} - \vec{\nabla} V \\ \vec{B} = \vec{\nabla} \wedge \vec{A} \end{cases} \quad [1.100]$$

In what follows, we will use the simplified expressions of the electric field $\vec{E}$ and the magnetic field $\vec{B}$ by omitting point M and time t.

1.1.3.4. *Electromotive field, induced electromotive force*

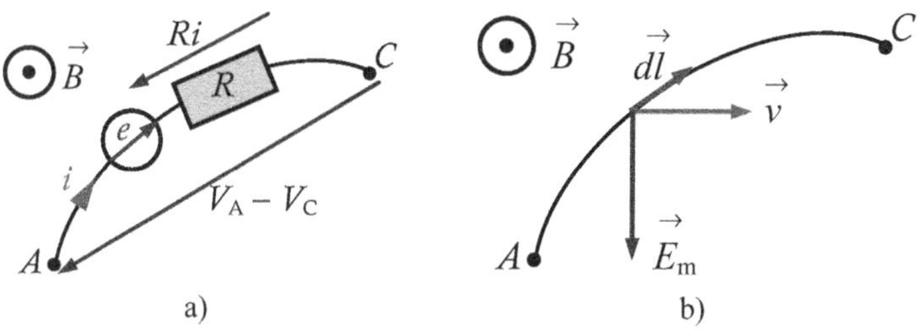

Figure 1.17. *a) Portion AC of a conductor in motion in a magnetic field; b) electric circuit equivalent to conductor A*

Let us consider AC as a portion of a dimensionally stable filiform conductor in motion with velocity $\vec{v}$ in a magnetic field $\vec{B}$ that can vary with time (Figure 1.17a).

By definition, the electromotive field $\vec{E}_m$ is given by:

$$\vec{E}_m = -\frac{\partial \vec{A}}{\partial t} + \vec{v} \wedge \vec{B} \quad [1.101]$$

The electromotive force induced in the portion of the circuit AC is equal to the circulation of the electromotive force between points A and C:

$$e = \int_A^C \vec{E}_m \cdot \vec{dl} \quad [1.102]$$

The potential difference between A and C is equal to:

$$V_A - V_C = Ri - e \quad [1.103]$$

1.1.3.5. *Maxwell-Faraday and Lenz-Faraday laws*

Another definition of the electromotive force can be obtained from the electromagnetic force [1.86]. An electron of conductor AC is excited by the overall motion at velocity $\vec{v}_e$ and is driven by the movement of the conductor at velocity $\vec{v}_C$. The overall velocity of an electron in the magnetic field is therefore $\vec{v} = \vec{v}_C + \vec{v}_e$. The electromagnetic force [1.86] acting on an electron of the conductor is therefore:

$$\vec{f} = q(\vec{E} + \vec{v} \wedge \vec{B}) = q(\vec{E} + (\vec{v}_C + \vec{v}_e) \wedge \vec{B}) = q(\vec{E} + \vec{v}_C \wedge \vec{B} + \vec{v}_e \wedge \vec{B}) \quad [1.104]$$

By definition, the electromotive force induced is equal to the circulation of force $\vec{f}/q$. So using [1.104]:

$$e = \int_{(C)} \frac{\vec{f}}{q} \cdot \vec{dl} = \int_{(C)} (\vec{E} + \vec{v}_C \wedge \vec{B} + \vec{v}_e \wedge \vec{B}) \cdot \vec{dl} \quad [1.105]$$

Considering Figure 1.17a, we see that the element with length $\vec{dl}$ has the same direction as the velocity $\vec{v}_e$ of the movement of all electrons. Therefore $\vec{v}_e \wedge \vec{dl} = \vec{0}$. Hence [1.105] gives:

$$e = \int_{(C)} (\vec{E} + \vec{v}_C \wedge \vec{B}) \cdot \vec{dl} \quad [1.106]$$

Equation [1.106] draws on two scenarios.

1st scenario: the conductor AC is mobile in a constant magnetic field

Considering [1.106], it becomes:

$$e = \int_{(C)} \vec{E} \cdot \vec{dl} + \int_{(C)} (\vec{v}_C \wedge \vec{B}) \cdot \vec{dl} \quad [1.107]$$

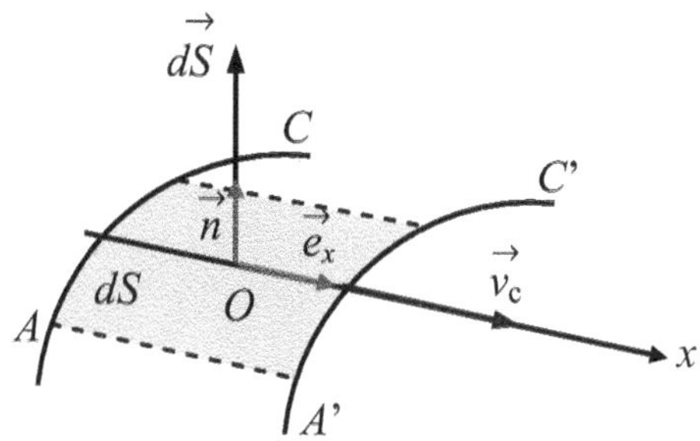

Figure 1.18. *Area swept by a section* AC *of the filiform conductor in motion*

For induction phenomena independent of time, the rotational of the electric field is zero according to [1.12]. Using Stokes' theorem, the first integral of the right term in equation [1.107] is transformed as follows:

$$\int_{(C)} \vec{E}\cdot\vec{dl} = \iint_{(S)} (\vec{\nabla}\wedge\vec{E})\cdot\vec{dS} = 0$$

The induced electromotive force [1.107] is written:

$$e = \int_{(C)} (\vec{v}_C \wedge \vec{B})\cdot\vec{dl} \qquad [1.108]$$

Using Stokes' theorem, the integral [1.108] is transformed as follows:

$$e = \iint_{(S)} \vec{\nabla}\wedge(\vec{v}_C \wedge \vec{B})\cdot\vec{dS} \qquad [1.109]$$

To transform the vector double product in equation [1.109], we consider the following property:

$$\vec{g}\wedge\vec{h}\wedge\vec{k} = \vec{h}\cdot(\vec{g}\cdot\vec{k}) - \vec{k}\cdot(\vec{g}\cdot\vec{h})$$

Moreover, regardless of the considered current distributions, the divergence of the magnetic field is zero [1.59]. Equation [1.109] is written as:

$$e = \iint_{(S)} \left[\vec{v}_C\cdot(\vec{\nabla}\cdot\vec{B}) - \vec{B}\cdot(\vec{\nabla}\cdot\vec{v}_C)\right]\cdot\vec{dS} = -\iint_{(S)} \vec{B}\cdot(\vec{\nabla}\cdot\vec{v}_C)\cdot\vec{dS}$$

Considering Figure 1.18, the last term in the above equation is written as follows:

$$e = -\iint_{(S)} \vec{B} \cdot (\vec{\nabla} \cdot \vec{v}_c) \cdot \overrightarrow{dS} = -\iint_{(S)} (\vec{\nabla} \cdot \vec{v}_c) \cdot (\vec{B} \cdot \overrightarrow{dS})$$

Therefore:

$$e = -\iint_{(S)} \left(\frac{d}{dx} \vec{e}_x \cdot \frac{dx}{dt} \vec{e}_x \right) \cdot (\vec{B} \cdot \overrightarrow{dS})$$

So, after simplification and rearrangement:

$$e = -\iint_{(S)} \vec{B} \cdot \overrightarrow{dS} = -\left(\frac{d\Phi}{dt} \right)_{mov} \qquad [1.110]$$

The result [1.110] is the *first law of electromagnetic induction*: the electromotive force induced is equal to the opposite of the variation in magnetic flux across any surface (*S*) based on the circuit (*C*). The variation in flux is due to the variation in surface area swept by the moving conductor. This justifies the index *mov* in [1.110].

2nd scenario: the conductor *AC* is fixed in a magnetic field variable with time

Considering [1.106], using Stokes' theorem, it becomes:

$$e = \int_{(C)} \vec{E} \cdot \vec{dl} = \iint_{(S)} (\vec{\nabla} \wedge \vec{E}) \cdot \overrightarrow{dS} \qquad [1.111]$$

For time-dependent phenomena, we are more in the realm of electrostatics. Consequently, the rotational of the electric field is therefore not zero. The electric field is given by the first of equations [1.100]:

$$\vec{E} = -\frac{\partial \vec{A}}{\partial t} - \vec{\nabla} V \qquad [1.112]$$

The rotational of the electric field is written as:

$$\vec{\nabla} \wedge \vec{E} = -\frac{\partial \vec{\nabla} \wedge \vec{A}}{\partial t} - \vec{\nabla} \wedge \vec{\nabla} V$$

Knowing that $\vec{\nabla} \wedge \vec{\nabla}V = \vec{0}$ and $\vec{B} = \vec{\nabla} \wedge \vec{A}$, we find:

$$\vec{\nabla} \wedge \vec{E} = -\frac{\partial \vec{B}}{\partial t} \qquad [1.113]$$

Equation [1.113] is obtained by Maxwell and presented at the local level, Faraday's experimental result on electromagnetic induction is known as the *Maxwell-Faraday law.*

Using [1.112], we express the electromotive force [1.111] as a function of magnetic flux.

We successively obtain:

$$e = \iint_{(S)} \vec{\nabla} \wedge \left(-\frac{\partial \vec{A}}{\partial t} - \vec{\nabla}V \right) \cdot \vec{dS} = -\iint_{(S)} \frac{\partial \vec{\nabla} \wedge \vec{A}}{\partial t} \cdot \vec{dS} - \iint_{(S)} \vec{\nabla} \wedge \vec{\nabla}V \cdot \vec{dS}$$

That is to say:

$$e = -\iint_{(S)} \frac{\partial \vec{\nabla} \wedge \vec{A}}{\partial t} \cdot \vec{dS} = -\iint_{(S)} \frac{\partial \vec{B}}{\partial t} \cdot \vec{dS}$$

And ultimately:

$$e = -\frac{\partial}{\partial t} \iint_{(S)} \vec{B} \cdot \vec{dS} = -\left(\frac{\partial \Phi}{\partial t} \right)_{magn} \qquad [1.114]$$

Result [1.114] is the *second law of electromagnetic induction*: the electromotive force induced is equal to the opposite of the variation in magnetic flux across a surface (*S*) based on the circuit (*C*). The variation in flux is due to the variation in the magnetic field over time. This justifies the index *magn*.

General case

Now let us generalize results [1.110] and [1.114] to any circuit (*C*) in motion in a magnetic field variable with time. The electromotive force induced is given by the equation:

$$e = \int_{(C)} (\vec{E} + \vec{v}_c \wedge \vec{B}) \cdot \vec{dl} = -\left(\frac{d\Phi}{dt} \right)_{mov} - \left(\frac{d\Phi}{dt} \right)_{magn} \qquad [1.115]$$

So, if:

$$\left(\frac{d\Phi}{dt}\right)_{total} = \left(\frac{d\Phi}{dt}\right)_{mov} + \left(\frac{d\Phi}{dt}\right)_{magn}$$

we obtain:

$$e = -\left(\frac{d\Phi}{dt}\right)_{total} \qquad [1.116]$$

Equation [1.116] reflects the *Lenz-Faraday law*. The induction phenomenon only appears if there is a variation in magnetic flux over time across the considered surface. In practice, we refer to fundamental equations [1.110] and [1.114] according to the nature of the problem studied.

We can therefore choose to fix the conductor and vary the magnetic field over time or consider a conductor in motion in a constant magnetic field.

1.1.3.6. *Lorentz and Neumann inductions*

Considering equation [1.115], it becomes:

$$e = -\left(\frac{d\Phi}{dt}\right)_{mov} - \left(\frac{d\Phi}{dt}\right)_{magn} \qquad [1.117]$$

Equation [1.117] highlights two types of electromagnetic induction: induction according to Lorentz and induction according to Neumann.

Lorentz induction

In the case of Lorentz induction, the conductor is mobile in a stationary (constant) magnetic field. The vector potential $\vec{A}$ is therefore constant and the electromotive field and induced electromotive force are written according to [1.101] and [1.102] as follows:

$$\overrightarrow{E_m} = \vec{v} \wedge \vec{B} \ ; e = \int_A^C \overrightarrow{E_m} \cdot \overrightarrow{dl} = \int_A^C (\vec{v} \wedge \vec{B}) \cdot \overrightarrow{dl} \qquad [1.118]$$

We find the expression [1.108] of the induced electromotive force by changing $\vec{v}$ into $\vec{v}_C$. This gives [1.110] which we recall below:

$$e = -\iint_{(S)} \vec{B} \cdot \overrightarrow{dS} = -\left(\frac{d\Phi}{dt}\right)_{mouv}$$

So, for a stationary magnetic field, the electromagnetic induction phenomena due to the variation in magnetic flux induced by the variation in surface swept by a conductor in motion is Lorentz induction.

Neumann induction

In the case of Neumann induction, the conductor considered is immobile in a time-dependent magnetic field. The vector potential $\vec{A}$ is therefore a function of time. Using equations [1.101] and [1.102], we express the electromotive field and the induced electromotive force.

We therefore obtain:

$$\overrightarrow{E_m} = -\frac{\partial \vec{A}}{\partial t}; \; e = \int_A^C \overrightarrow{E_m} \cdot \overrightarrow{dl} = -\int_A^C \frac{\partial \vec{A}}{\partial t} \cdot \overrightarrow{dl} \qquad [1.119]$$

We find expression [1.111] of the induced electromotive force. This gives the following result [1.114]:

$$e = -\frac{\partial}{\partial t} \iint_{(S)} \vec{B} \cdot \overrightarrow{dS} = -\left(\frac{\partial \Phi}{\partial t}\right)_{magn}$$

For a fixed conductor, electromagnetic induction phenomena due to the variation of magnetic flux induced by the variation in the magnetic field is Neumann induction.

Franz Ernst Neumann was an American mathematician, physicist and mineralogist. In the field of physics, he is most famous for his work on induction. He demonstrated before Helmholtz (1797–1878) that the phenomenon of electromagnetic induction can be studied via the conservation of energy. Working on electrodynamics, Neumann published the laws of induction between 1845 and 1847. Starting with work by Ampère (1775–1836), Neumann introduced in 1845 the important concept of "mutual potential" between two electrical circuits. The phenomenon of electromagnetic induction introducing the coefficients of mutual inductance between two electrical circuits is also called Neumann induction.

Box 1.9. *Neumann (1798–1895)*

1.1.3.7. *Coulomb's law*

Here we refer to an electric conductor in equilibrium. Initially, the conductor is neutral (Figure 1.19a) then uniformly charged on the surface (Figure 1.19b). The surface charge density chosen is positive and denoted σ. The *equilibrium condition* of a conductor implies the immobility of its internal charges.

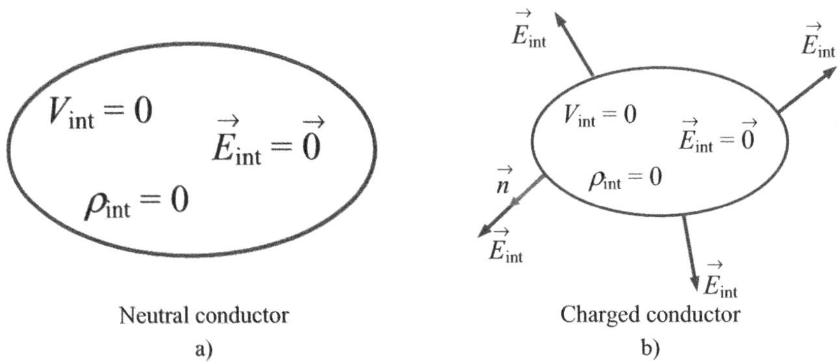

Figure 1.19. *Conductors in equilibrium*

In a neutral conductor in equilibrium, the surface charge density or the volume charge density is zero, the electrostatic field is zero and the potential is constant (Figure 1.19a), i.e.:

$$\sigma = 0;\ \rho_{int} = 0;\ E_{int} = 0;\ V_{int} = C^{te}$$

In a charged conductor in equilibrium, the charges can only be distributed across the surface: the surface charge density is therefore non-zero, the volume charge density is zero, the electrostatic field is zero and the electrostatic potential is constant (Figure 1.19b) i.e.:

$$\sigma \neq 0;\ \rho_{int} = 0;\ E_{int} = 0;\ V_{int} = C^{te}$$

In the vicinity of the surface of a charged conductor, the external electrostatic field is normal to the surface of the conductor and equal to:

$$\vec{E}_{ext} = \frac{\sigma}{\varepsilon_0}\vec{n} \qquad [1.120]$$

In this equation, the normal unit vector $\vec{n}$ is perpendicular to the surface and orientated towards the exterior of the conductor (the surface of the conductor is closed).

– If $\sigma > 0$, the electrostatic field is directed towards the exterior of the conductor (Figure 1.19b).

– If $\sigma < 0$, the electrostatic field is directed towards the interior of the conductor.

Result [1.120] reflects *Coulomb's law*. According to this law, *the electrostatic field lines of a charged conductor in equilibrium are normal to the surface of the conductor.*

Let us demonstrate Coulomb's law [1.120] before continuing. If we consider an infinite flat surface, uniformly charged with a surface charge density $\sigma > 0$ (Figure 1.20). To apply Gauss's theorem, we consider a cylinder with base surface S and height h. The total flux of the electrostatic field exiting the cylinder from the base surfaces dS and dS' when the field changes from P to P' is:

$$\Phi = \iint_{(S)} \left[\vec{E}(z) \cdot \overrightarrow{dS} + \vec{E}(-z) \cdot \overrightarrow{dS}' \right] = \frac{Q_{\text{int}}}{\varepsilon_0} \qquad [1.121]$$

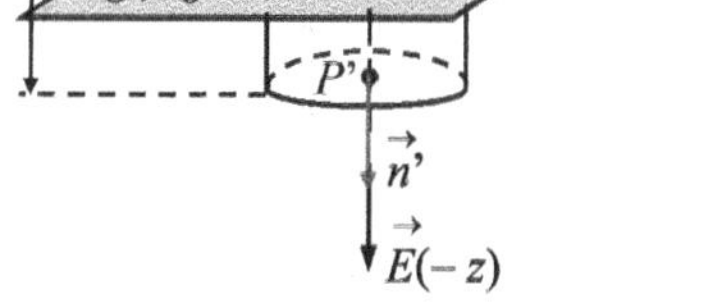

Figure 1.20. *Electrostatic field surrounding a flat charged surface*

In addition:

$dS = dS'$; $E(z) = |E(-z)|$; $Q_{\text{int}} = \sigma S$

Hence, according to [1.121], the flux of the electrostatic field is:

$$\Phi = 2ES = \sigma S/\varepsilon_0$$

Ultimately, we obtain:

$$E = \frac{\sigma}{2\varepsilon_0} \qquad [1.122]$$

Using result [1.122], the electrostatic field at points P and P' is given by the respective equations:

$$\begin{cases} \vec{E}(P) = \dfrac{\sigma}{2\varepsilon_0}\vec{n} \\ \vec{E}(P') = -\dfrac{\sigma}{2\varepsilon_0}\vec{n} \end{cases} \qquad [1.123]$$

When crossing the charged surface, the variation in the electrostatic field is written as:

$$\overrightarrow{\Delta E} = \vec{E}(P) - \vec{E}(P') = \frac{\sigma}{2\varepsilon_0}\vec{n} + \frac{\sigma}{2\varepsilon_0}\vec{n} = \frac{\sigma}{\varepsilon_0}\vec{n}$$

At point M located on the surface of the conductor, the electrostatic field is:

$$\vec{E}(M) = \frac{\sigma(M)}{\varepsilon_0}\vec{n} \qquad [1.124]$$

Coulomb's law is therefore demonstrated. The physical interpretation of this theorem is clear: *when crossing a charged surface, the normal component of the electrostatic field undergoes a discontinuity proportional to the surface charge density* σ. The coefficient of proportionality is equal to $1/\varepsilon_0$.

1.1.3.8. *Electrical energy density*

Let us consider a set Σ of n conductors C_1, C_2, C_3, …, C_i, …, C_n in equilibrium (Figure 1.21).

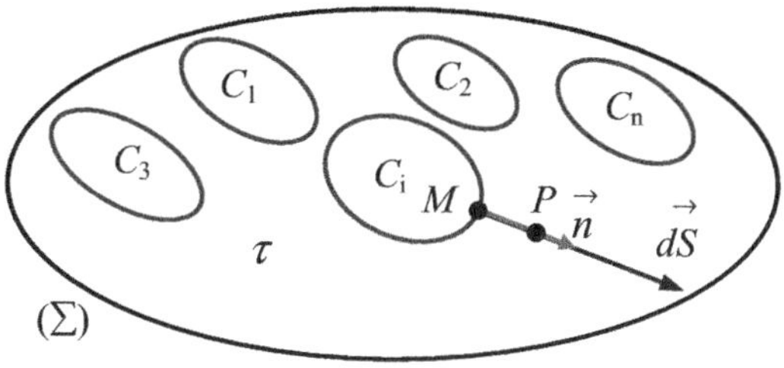

Figure 1.21. *Conductors in equilibrium*

The electrostatic field at point M is given by [1.124]. For a continuous charge distribution, the potential energy W_p is written as:

$$W_p = \frac{1}{2}\int V dq$$

At a point on the surface of the conductor i situated in volume V, we obtain:

$$W_p = \frac{1}{2}\iint_{(S)} V dq = \frac{1}{2}\iint_{(S)} V \sigma dS \qquad [1.125]$$

If we take σ from expression [1.124] and then together with the result obtained in the above equation, it becomes:

$$W_p = \iint_{(S)} \frac{\varepsilon_0}{2}[V\vec{E}\cdot\vec{n}]dS = \iint_{(S)} \frac{\varepsilon_0}{2}[V\vec{E}]\cdot\vec{n}\,dS$$

So, by introducing the surface vector:

$$W_p = \iint_{(S)} \frac{\varepsilon_0}{2}[V\vec{E}]\cdot\overrightarrow{dS} \qquad [1.126]$$

Using the divergence theorem, we obtain from [1.126]:

$$W_p = -\iiint_{(\tau)} \frac{\varepsilon_0}{2}\vec{\nabla}\cdot[V\vec{E}]d\tau \qquad [1.127]$$

The sign "–" in expression [1.127] is due to the fact that the unit vector $\vec{n}$ normal to the surface dS is directed inside the volume τ (Figure 1.21).

Let us take advantage of the vector analysis equation verified by any scalar function f:

$$\vec{\nabla}\cdot(f\vec{g}) = f\vec{\nabla}\cdot\vec{g} + \vec{g}\cdot\vec{\nabla}f$$

Using this equation and considering the local law $\vec{E} = -\vec{\nabla}V$, the potential energy [1.127] is transformed as follows:

$$W_p = -\iiint_{(\tau)} \frac{1}{2\varepsilon_0}\left[V\vec{\nabla}\cdot\vec{E} + \vec{E}\cdot\vec{\nabla}V\right]d\tau$$

That is to say:

$$W_p = -\iiint_{(\tau)} \frac{1}{2\varepsilon_0}\left[V\vec{\nabla}\cdot\vec{E} - \vec{E}^2\right]d\tau$$

With the conductor in equilibrium, the volume charge density $\rho = 0$. According to the local form of [1.164] Gauss's theorem, $\vec{\nabla}\cdot\vec{E} = 0$. This gives:

$$W_p = \iiint_{(\tau)} \frac{1}{2}\varepsilon_0 \vec{E}^2 d\tau = \iiint_{(\tau)} \frac{1}{2}\varepsilon_0 E^2 d\tau \qquad [1.128]$$

Considering equation [1.128], the potential energy W_p appears to be distributed throughout the whole volume τ of space between the conductors with an electrical energy density w_e is equal to (omitting point M):

$$w_e = \frac{dW_p}{dV} = \frac{1}{2}\varepsilon_0 E^2 \qquad [1.129]$$

1.1.3.9. *Magnetic volume energy density*

This section concerns the distribution of conductors in equilibrium as shown in Figure 1.21. Volume τ is the site of a current distribution in steady state. Let's denote $\vec{A}(M)$ and $\vec{j}(M)$ the vector potential and the current density vector at point M respectively. Using [1.70], we obtain:

$$\vec{A}(M) = \iiint_{(V)} \frac{\mu_0 \vec{j}(M) dV}{4\pi r} \quad [1.130]$$

We then proceed by analogy to establish the magnetic volume energy density. For a charge volume density distribution ρ, the potential energy W_p of the system of charges is given by the equation:

$$W_p = \frac{1}{2} \iiint_{(\tau)} V dq = \frac{1}{2} \iiint_{(\tau)} V \rho d\tau \quad [1.131]$$

By analogy, the potential energy W_p of the current distribution in volume τ is deduced from expression [1.130] by replacing the scalar potential with the vector potential and the volume charge density with the current density vector.

Therefore:

$$W_p = \frac{1}{2} \iiint_{(\tau)} \vec{A} \vec{J} dV \quad [1.132]$$

Now in [1.132] we need to introduce the square of the magnetic field as in expression [1.128]. According to [1.25] and [1.26] we know that:

$$\vec{B} = \vec{\nabla} \wedge \vec{A} \text{ and } \vec{\nabla} \wedge \vec{B} = \mu_0 \vec{J}$$

Let us consider the transformation equation for the divergence of the vector product of the vector potential and the magnetic field:

$$\vec{\nabla} \cdot (\vec{A} \wedge \vec{B}) = \vec{B} \cdot \vec{\nabla} \wedge \vec{A} - \vec{A} \cdot \vec{\nabla} \wedge \vec{B} = \vec{B}^2 - \mu_0 \vec{A} \cdot \vec{J}$$

This gives:

$$\mu_0 \vec{A} \cdot \vec{J} = B^2 - \vec{\nabla} \cdot (\vec{A} \wedge \vec{B})$$

Using this result, expression [1.132] is written, omitting point M:

$$W_p = \iiint_{(\tau)} \frac{B^2}{2\mu_0} d\tau - \iiint_{(\tau)} \left[\vec{\nabla} \cdot (\vec{A} \wedge \vec{B}) \right] d\tau \quad [1.133]$$

Let us take into account the vector analysis equation verified by ordinary vectors:

$$\vec{f} \wedge (\vec{g} \wedge \vec{h}) = \vec{g} \cdot (\vec{f} \cdot \vec{h}) - \vec{h} \cdot (\vec{f} \cdot \vec{g}) = \vec{g} \cdot (\vec{f} \cdot \vec{h}) - (\vec{f} \cdot \vec{g}) \cdot \vec{h}$$

Using this result, knowing that $\vec{B} = \vec{\nabla} \wedge \vec{A}$, we obtain the following, according to [1.134]:

$$\vec{\nabla} \cdot (\vec{A} \wedge \vec{B}) = \vec{\nabla} \cdot \left[\vec{A} \wedge (\vec{\nabla} \wedge \vec{A}) \right] = \vec{\nabla} \cdot \left[\vec{\nabla} \cdot (\vec{A} \cdot \vec{A}) - \vec{A} \cdot (\vec{A} \cdot \vec{\nabla}) \right]$$

Given that the nabla vector is not an ordinary vector (it is an operator), it becomes:

$$\vec{\nabla} \cdot (\vec{A} \wedge \vec{B}) = \vec{\nabla} \cdot \left[\vec{\nabla} \cdot (\vec{A} \cdot \vec{A}) - (\vec{\nabla} \cdot \vec{A}) \cdot \vec{A} \right] = \vec{\nabla} \cdot (\vec{\nabla} A^2 - \vec{\nabla} A^2) = 0$$

With this result, the magnetic potential energy [1.133] is written:

$$W_p = \iiint_{(\tau)} \frac{B^2}{2\mu_0} d\tau \qquad [1.134]$$

The magnetic potential energy [1.134] also appears to be distributed throughout volume τ of space between the conductors with a magnetic volume energy density w_{m} equal to:

$$w_m = \frac{dW_p}{dV} = \frac{B^2}{2\mu_0} \qquad [1.135]$$

1.1.4. *Maxwell's equations*

1.1.4.1. *Correction of Ampère's law: displacement current*

Maxwell's equations link the values of the electric field $\vec{E} = \vec{E}(\vec{r},t)$ and the magnetic field $\vec{B} = \vec{B}(\vec{r},t)$ at infinitely neighboring points at infinitely neighboring dates. In a variable regime, $\dfrac{\partial \rho}{\partial t} \neq 0$. Hence, according to the continuity equation [1.51]:

$$\vec{\nabla} \cdot \vec{J} = -\frac{\partial \rho}{\partial t} \neq 0 \qquad [1.136a]$$

Using the local law [1.64] of Ampère's law, knowing that the divergence of a rotational is zero, we obtain:

$$\vec{\nabla} \wedge \vec{B} = \mu_0 \vec{J} \Rightarrow \vec{\nabla} \cdot (\vec{\nabla} \wedge \vec{B}) = \mu_0 \vec{\nabla} \cdot \vec{J} = 0 \qquad [1.136b]$$

Results [1.136a] and [1.136b] show that there is incompatibility between the charge conservation law [1.136a] and the local formulation of Ampère's law [1.136b]. As the charge conservation law is a fundamental law of physics that has never been disproven experimentally, Maxwell chose to modify Ampère's law to simplify it. He added the displacement current $\vec{j}_d$ to the current density vector, defined by the equation:

$$\vec{J}_d = \varepsilon_0 \frac{\partial \vec{E}}{\partial t} \qquad [1.136c]$$

Ampère's generalized law according to Maxwell is written:

$$\vec{\nabla} \wedge \vec{B} = \mu_0 \left[\vec{J} + \varepsilon_0 \frac{\partial \vec{E}}{\partial t} \right] = \mu_0 \vec{J} + \mu_0 \varepsilon_0 \frac{\partial \vec{E}}{\partial t} \qquad [1.136d]$$

The displacement current [1.136c] concerns a fictitious current introduced to correct Ampère's law so that it is compatible with the law of conservation of the electric charge in variable regime. However, the introduction of the displacement current into Ampère's generalized theorem gives rise to all phenomena relating to the propagation of electromagnetic waves in space. Thus, Maxwell's equations predict the propagation of electromagnetic waves.

NOTE.– We can verify that Ampère's generalized law [1.136] satisfies the charge conservation principle. Indeed, the divergence of a rotational is zero, so:

$$\vec{\nabla} \cdot (\vec{\nabla} \wedge \vec{B}) = \vec{\nabla} \cdot \left[\mu_0 \vec{J} + \mu_0 \varepsilon_0 \frac{\partial \vec{E}}{\partial t} \right] = 0 \Rightarrow \mu_0 \vec{\nabla} \cdot \vec{J} + \mu_0 \varepsilon_0 \frac{\partial \vec{\nabla} \cdot \vec{E}}{\partial t} = 0$$

Considering the local formulation [1.25] Gauss's theorem is:

$$\vec{\nabla} \cdot \vec{E} = \rho / \varepsilon_0$$

Whereby:

$$\vec{\nabla}\cdot\vec{J}+\varepsilon_0\frac{\partial\vec{\nabla}\cdot\vec{E}}{\partial t}=0\Rightarrow\vec{\nabla}\cdot\vec{J}+\frac{\partial\rho}{\partial t}=0$$

1.1.4.2. *Maxwell's equation couples*

Maxwell's equations are grouped into equation couples:

– the *first couple* corresponds to equations deduced from the definition of the electromagnetic field: *Maxwell-Faraday equation* and *Maxwell-flux equation*;

– the *second couple* corresponds to equations concerning the local formulation of Gauss's theorem and the generalization of Ampère's law: *Maxwell-Gauss equation* and *Maxwell-Ampère equation.*

First Maxwell equation couple

In a region of space based on a Galilean frame of reference where a vector potential $\vec{A}=\vec{A}(\vec{r},t)$ and a scalar potential $V=V(r,t)$ coexistent, an electromagnetic field occurs. This field is characterized by the electric field $\vec{E}=\vec{E}(\vec{r},t)$ and the magnetic field $\vec{B}=\vec{B}(\vec{r},t)$ given by expressions [1.100] that we remember below:

$$\vec{E}=-\frac{\partial\vec{A}}{\partial t}-\vec{\nabla}V\ ;\ \vec{B}=\vec{\nabla}\wedge\vec{A}$$

Since the rotational of a gradient and the divergence of a rotational are zero, then:

$$\vec{\nabla}\wedge\vec{E}=-\frac{\partial\vec{\nabla}\wedge\vec{A}}{\partial t}-\vec{\nabla}\wedge\vec{\nabla}V=-\frac{\partial\vec{B}}{\partial t};\ \vec{\nabla}\cdot\vec{B}=\vec{\nabla}\cdot(\vec{\nabla}\wedge\vec{A})=0$$

Hence the *first Maxwell equation couple* is:

$$\vec{\nabla}\wedge\vec{E}=-\frac{\partial\vec{B}}{\partial t}\ \text{and}\ \vec{\nabla}\cdot\vec{B}=0 \qquad [1.137]$$

Second Maxwell equation couple

As we have specified previously, the second couple of Maxwell equations summarizes the local formulation of Gauss's theorem and the generalization of Ampère's law, i.e.:

$$\vec{\nabla} \wedge \frac{\vec{B}}{\mu_0} = \vec{J} + \varepsilon_0 \frac{\partial \vec{E}}{\partial t} \text{ and } \vec{\nabla} \cdot \vec{E} = \frac{\rho}{\varepsilon_0} \quad [1.138]$$

In the empty space of charges and currents, the electromagnetic waves propagate at the speed of light c such that:

$$\mu_0 \varepsilon_0 c^2 = 1 \Rightarrow c = \frac{1}{\sqrt{\mu_0 \varepsilon_0}} \quad [1.139]$$

In the international system:

$$\mu_0 = 4\pi \times 10^{-7} \text{ H} \cdot \text{m}^{-1};\ \varepsilon_0 = (1/36\pi 10^9) \text{ F} \cdot \text{m}^{-1};\ c = 3 \times 10^{8} \text{ m} \cdot \text{s}^{-1}$$

In Table 1.4, we recap Maxwell's equations which constitute the *fundamental equations of electromagnetism* in a vacuum.

First couple	$\vec{\nabla} \wedge \vec{E} = -\frac{\partial \vec{B}}{\partial t}$	Maxwell-Faraday
	$\vec{\nabla} \cdot \vec{B} = 0$	Maxwell-flux
Second couple	$\vec{\nabla} \wedge \frac{\vec{B}}{\mu_0} = \vec{j} + \varepsilon_0 \frac{\partial \vec{E}}{\partial t}$	Maxwell-Ampère
	$\vec{\nabla} \cdot \vec{E} = \frac{\rho}{\varepsilon_0}$	Maxwell-Gauss

Table 1.4. *The four Maxwell fundamental equations in a vacuum*

With regard to Table 1.4, we notice that the Maxwell-flux and Maxwell-Gauss equations are asymmetrical. The divergence of the magnetic field is zero: the magnetic field is of conservative flux; which proves the absence of magnetic charges or magnetic monopoles. However, the fact that the divergence of the electric field is not zero conveys the existence of electric charges (the flow of the electric field is linked to all of the internal source charges of the field) or electric monopoles.

1.1.4.3. *Principle of superposition*

The Maxwell equations in Table 1.4 are linear with respect to divergence operators, the rotational and the temporal partial derivative. We designate $\rho = \rho(\vec{r},t)$ and $\vec{J} = \vec{J}(\vec{r},t)$ the field sources creating the electric field $\vec{E} = \vec{E}(\vec{r},t)$ and the magnetic field $\vec{B} = \vec{B}(\vec{r},t)$. Similarly, we denote $\rho_1 = \rho_1(\vec{r},t)$ and $\vec{J}_1 = \vec{J}_1(\vec{r},t)$ the field sources creating the electric $\vec{E}_1 = \vec{E}_1(\vec{r},t)$ and the magnetic $\vec{B}_1 = \vec{B}_1(\vec{r},t)$ fields and we denote $\rho_2 = \rho_2(\vec{r},t)$ and $\vec{J}_2 = \vec{J}_2(\vec{r},t)$ the field sources creating the electric $\vec{E}_2 = \vec{E}_2(\vec{r},t)$ and the magnetic $\vec{B}_2 = \vec{B}_2(\vec{r},t)$ fields.

The resulting field sources and the electric and magnetic fields satisfy the principle of superposition. We obtain the following equations [1.140]:

$$\begin{cases} \rho = \lambda_1 \rho_1 + \lambda_2 \rho_2 \\ \vec{J} = \lambda_1 \vec{J}_1 + \lambda_2 \vec{J}_2 \end{cases} \Rightarrow \begin{cases} \vec{E} = \lambda_1 \vec{E}_1 + \lambda_2 \vec{E}_2 \\ \vec{B} = \lambda_1 \vec{B}_1 + \lambda_2 \vec{B}_2 \end{cases} \qquad [1.140]$$

Note that the integration with respect to time of the differential equations, obtained using Maxwell's equations to determine the electric $\vec{E}$ and magnetic $\vec{B}$ fields, naturally gives rise to integration constants. These constants can be interpreted as static fields (electrostatic field and magnetostatic field). However, if the phenomena studied depend exclusively on time, these integration constants must be zero since they do not affect the phenomena studied.

James Clerk Maxwell was a Scottish physicist. His fame is linked to the Maxwell distribution in statistical physics but above all to his important contributions in electromagnetism. The so-called Maxwell fundamental equations, established in 1864, unite electrostatics, magnetostatics and induction. Maxwell also owes his fame to the interpretation of light as an electromagnetic phenomenon based on the work of Faraday (1791–1867). Maxwell demonstrated in particular that the electric and magnetic fields propagate in space in the form of light waves (electromagnetic waves) at the speed of light.

Box 1.10. *Maxwell (1831–1879)*

1.1.4.4. *Invariance of the electromagnetic field*

In a given frame of reference, the electromagnetic force [1.86] and the electric and magnetic fields [1.100] are individually determined. In other words, at each point M in space and at any instant, there is a single electric field $\vec{E}$ and a single magnetic field $\vec{B}$. However, this exclusivity of the electric and magnetic fields is not verified by the scalar V and vector $\vec{A}$ potentials, as we will demonstrate.

Here we have two couples of potentials ($\vec{A}$, V) and ($\vec{A}^*$, V^*). We obtain both from the following figure:

– the magnetic field $\vec{B}$ being unique, is:

$$\vec{B} = \vec{\nabla} \wedge \vec{A} = \vec{\nabla} \wedge \vec{A}^* \Rightarrow \vec{\nabla} \wedge (\vec{A}^* - \vec{A}) = \vec{0}$$

Given that the rotational of the gradient of a scalar function is zero, considering the scalar function φ, we obtain:

$$\vec{A}^* = \vec{A} + \vec{\nabla}\varphi \qquad [1.141]$$

– the electric field $\vec{E}$ being unique, using [1.100], it becomes:

$$\vec{E} = -\frac{\partial \vec{A}}{\partial t} - \vec{\nabla}V = -\frac{\partial \vec{A}^*}{\partial t} - \vec{\nabla}V^* \Rightarrow \frac{\partial(\vec{A}^* - \vec{A})}{\partial t} + \vec{\nabla}(V^* - V) = 0$$

So, by referring to [1.141], we obtain:

$$\frac{\partial}{\partial t}\vec{\nabla}\varphi + \vec{\nabla}(V^* - V) = 0 \Rightarrow \vec{\nabla}\left(\frac{\partial \varphi}{\partial t} + (V^* - V)\right) = 0$$

This gives the equation between the scalar potentials V^* and V:

$$V^* = V - \frac{\partial \varphi}{\partial t} \qquad [1.142]$$

Since the rotational of the gradient of a scalar function is zero, considering the scalar function φ, we obtain:

$$\vec{A}^* = \vec{A} + \vec{\nabla}\varphi \qquad [1.143]$$

CONCLUSION.– If an electromagnetic field corresponds to the couples of potentials ($\vec{A}$, V), at a given point in space and at a given instant, the electromagnetic field remains unchanged if the vector potential changes $\vec{A}$ into $\vec{A}^*$ and the scalar potential changes V into V^* such that:

$$\vec{A}^* = \vec{A} + \vec{\nabla}\varphi \text{ and } V^* = V - \frac{\partial \varphi}{\partial t}$$

In [1.142] and [1.143], φ is an arbitrary scalar function.

1.1.4.5. *Coulomb and Lorentz gauges*

Equations [1.142] show that there are several scalar potentials and several vector potentials that can generate the same electromagnetic field [1.100]. In practice, the most appropriate vector and/or scalar potential must be chosen for the problem studied. Usually, the potential chosen is the one that allows the calculations to be simplified as much as possible. For this, we impose conditions on the vector potential $\vec{A}$ and the scalar potential V. *Choosing the potential* best adapted to the problem studied is known as *gauge fixing*.

Let us consider the Maxwell-Gauss and Maxwell-Ampère equations from Table 1.4:

$$\vec{\nabla}\cdot\vec{E} = \frac{\rho}{\varepsilon_0} \text{ and } \vec{\nabla}\wedge\frac{\vec{B}}{\mu_0} = \vec{J} + \varepsilon_0\frac{\partial \vec{E}}{\partial t} \qquad [1.144]$$

Given that:

$$\vec{E} = \frac{\partial \vec{A}}{\partial t} - \vec{\nabla}V \text{ and } \vec{B} = \vec{\nabla}\wedge\vec{A}$$

it becomes:

$$\vec{\nabla}\cdot\vec{E} = -\vec{\nabla}\cdot\left(\frac{\partial \vec{A}}{\partial t} + \vec{\nabla}V\right) = \frac{\rho}{\varepsilon_0}$$

$$\vec{\nabla}\wedge\vec{B} = \vec{\nabla}\wedge\vec{\nabla}\wedge\vec{A} = \mu_0\vec{J} - \mu_0\varepsilon_0\frac{\partial}{\partial t}\left(\frac{\partial \vec{A}}{\partial t} + \vec{\nabla}V\right)$$

So:

$$\frac{\partial \vec{\nabla} \cdot \vec{A}}{\partial t} + \vec{\nabla} \cdot \vec{\nabla} V = -\frac{\rho}{\varepsilon_0} \text{ and } \vec{\nabla} \wedge \vec{\nabla} \wedge \vec{A} = \mu_0 \vec{J} - \mu_0 \varepsilon_0 \frac{\partial}{\partial t}\left(\frac{\partial \vec{A}}{\partial t} + \vec{\nabla} V\right)$$

Considering the vector identity equations:

$$\vec{\nabla} \cdot \vec{\nabla} f = \Delta f \text{ and } \vec{\nabla} \wedge \vec{\nabla} \wedge \vec{f} = \vec{\nabla}(\vec{\nabla} \cdot \vec{f}) - \Delta \vec{f}$$

We therefore obtain:

$$\frac{\partial \vec{\nabla} \cdot \vec{A}}{\partial t} + \Delta V = -\frac{\rho}{\varepsilon_0} \text{ and } \vec{\nabla}(\vec{\nabla} \cdot \vec{A}) - \Delta \vec{A} = \mu_0 \vec{J} - \mu_0 \varepsilon_0 \frac{\partial}{\partial t}\left(\frac{\partial \vec{A}}{\partial t} + \vec{\nabla} V\right)$$

Let us arrange the last equality in the equations above. It becomes:

$$\vec{\nabla}\left(\vec{\nabla} \cdot \vec{A} + \mu_0 \varepsilon_0 \frac{\partial V}{\partial t}\right) = \Delta \vec{A} - \mu_0 \varepsilon_0 \frac{\partial^2 \vec{A}}{\partial t^2} + \mu_0 \vec{J}$$

And ultimately:

$$\Delta V + \frac{\rho}{\varepsilon_0} = \frac{\partial \vec{\nabla} \cdot \vec{A}}{\partial t} \qquad [1.145a]$$

$$\Delta \vec{A} - \mu_0 \varepsilon_0 \frac{\partial^2 \vec{A}}{\partial t^2} + \mu_0 \vec{J} = \vec{\nabla}\left(\vec{\nabla} \cdot \vec{A} + \mu_0 \varepsilon_0 \frac{\partial V}{\partial t}\right) \qquad [1.145b]$$

Equation [1.145a] gives rise to Poisson's equation [1.66] if the divergence of the vector potential is zero. This is the first gauge condition imposed on the vector potential expressing the *Coulomb gauge* [1.68]:

$$\vec{\nabla} \cdot \vec{A} = 0$$

The Coulomb gauge [1.68] is valid in the field of magnetostatics and electrostatics since the potential V is independent of time. This leads to a time independent radial electric field. For time-dependent phenomena, the Coulomb gauge is no longer valid. It must be corrected by imposing an additional condition on the scalar potential. The simultaneous choosing of the vector potential and scalar potential is called the *Lorentz gauge*, expressed by the following conditions:

$$\begin{cases} \vec{\nabla} \cdot \vec{A} + \mu_0 \varepsilon_0 \dfrac{\partial V}{\partial t} = 0 \\ \vec{A}(\infty) = \vec{0}\,;\, V(\infty) = 0 \end{cases} \qquad [1.146]$$

Obtaining the divergence of the vector potential and inserting the result into equation [1.145a], and considering the Lorentz gauge [1.146], we obtain the analogous differential equations satisfied by the vector potential and the scalar potential:

$$\begin{cases} \Delta V - \mu_0 \varepsilon_0 \dfrac{\partial^2 V}{\partial t^2} + \dfrac{\rho}{\varepsilon_0} = 0 \\ \Delta \vec{A} - \mu_0 \varepsilon_0 \dfrac{\partial^2 \vec{A}}{\partial t^2} + \mu_0 \vec{j} = 0 \end{cases} \Rightarrow \begin{cases} \Delta V - \dfrac{1}{c^2} \dfrac{\partial^2 V}{\partial t^2} + \dfrac{\rho}{\varepsilon_0} = 0 \\ \Delta \vec{A} - \dfrac{1}{c^2} \dfrac{\partial^2 \vec{A}}{\partial t^2} + \mu_0 \vec{j} = 0 \end{cases} \qquad [1.147]$$

The solutions to the differential equations [1.147] are known as retarded potentials, given by the following expressions:

$$V(r,t) = \frac{1}{4\pi\varepsilon_0} \iiint_{(\tau')} \frac{\rho\left(\vec{r}', t - \left\|\vec{r} - \vec{r}'\right\|/c\right)}{\left\|\vec{r} - \vec{r}'\right\|} d\tau' \qquad [1.148a]$$

$$\vec{A}(\vec{r},t) = \frac{\mu_0}{4\pi} \iiint_{(\tau')} \frac{\vec{j}\left(\vec{r}', t - \left\|\vec{r} - \vec{r}'\right\|/c\right)}{\left\|\vec{r} - \vec{r}'\right\|} d\tau' \qquad [1.148b]$$

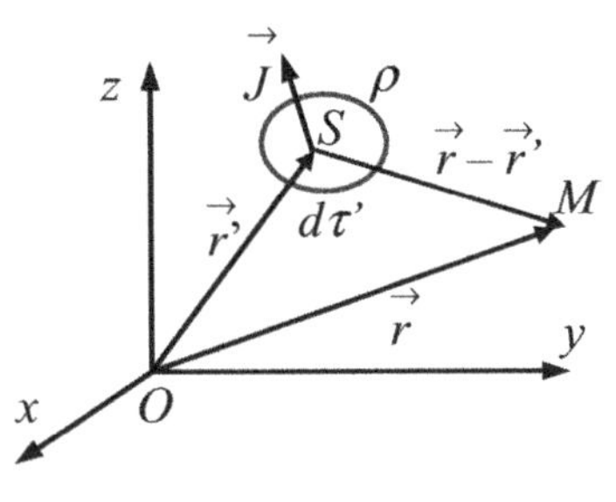

Figure 1.22. *Sources (S) of charges and currents*

We can justify the presence of the term ($t - c/r$) in expressions [1.148] where c is the propagation velocity of electromagnetic waves in an empty space of charges and currents. This term satisfies the *Principle of Causality* according to which the causes precede the effects.

In this case, the causes determined by the charge and current densities at source S precede the effects determined by the potentials created at point M (Figure 1.22). The creation of these potentials therefore creates a lag $\tau = r/c$ equal to the propagation time of waves from source S to point M. Thus, the scalar potential [1.148a] and the vector potential [1.148b] are created at time $t - \tau = (t - r/c)$.

1.1.4.6. *Propagation equations*

In the empty space of charges ($\rho = 0$) and currents ($\vec{J} = \vec{0}$), the Maxwell equations are written as follows (see Table 1.4):

$$\vec{\nabla} \wedge \vec{E} = -\frac{\partial \vec{B}}{\partial t}; \ \vec{\nabla} \cdot \vec{B} = 0; \ \vec{\nabla} \wedge \vec{B} = \mu_0 \varepsilon_0 \frac{\partial \vec{E}}{\partial t}; \ \vec{\nabla} \cdot \vec{E} = 0 \qquad [1.149]$$

The differential equation [1.147] satisfied by the vector potential is therefore written as:

$$\Delta\vec{A} - \frac{1}{c^2}\frac{\partial^2 \vec{A}}{\partial t^2} = 0 \qquad [1.150]$$

Using [1.150], it becomes:

$$\frac{\partial}{\partial t}\left[\Delta\vec{A} - \frac{1}{c^2}\frac{\partial^2 \vec{A}}{\partial t^2}\right] = \frac{\partial}{\partial t}(\Delta\vec{A}) - \frac{\partial}{\partial t}\left(\frac{1}{c^2}\frac{\partial^2 \vec{A}}{\partial t^2}\right) = 0$$

So, considering [1.149]:

$$\Delta\frac{\partial}{\partial t}\left(\vec{A}\right) - \frac{1}{c^2}\frac{\partial^2}{\partial t^2}\left(\frac{\partial \vec{A}}{\partial t}\right) = 0 \Rightarrow \Delta\vec{E} - \frac{1}{c^2}\frac{\partial^2 \vec{E}}{\partial t^2} = 0 \qquad [1.151a]$$

Also, for the magnetic field using [1.150], we find that:

$$\vec{\nabla} \wedge \left[\Delta\vec{A} - \frac{1}{c^2}\frac{\partial^2 \vec{A}}{\partial t^2}\right] = \vec{\nabla} \wedge (\Delta\vec{A}) - \vec{\nabla} \wedge \left(\frac{1}{c^2}\frac{\partial^2 \vec{A}}{\partial t^2}\right) = 0$$

So, considering [1.149]:

$$\Delta(\vec{\nabla}\wedge\vec{A})-\frac{1}{c^2}\frac{\partial^2}{\partial t^2}(\vec{\nabla}\wedge\vec{A})=0\Rightarrow\Delta\vec{B}-\frac{1}{c^2}\frac{\partial^2\vec{B}}{\partial t^2}=0 \qquad [1.151b]$$

Using [1.151], we obtain the propagation equations of the electric and magnetic fields that are analogous to the propagation equation [1.150] of the vector potential. Therefore:

$$\Delta\vec{E}-\frac{1}{c^2}\frac{\partial^2\vec{E}}{\partial t^2}=0\,;\ \Delta\vec{B}-\frac{1}{c^2}\frac{\partial^2\vec{B}}{\partial t^2}=0 \qquad [1.152]$$

The propagation equations [1.150] and [1.152] determine the equations for the evolution of an electromagnetic field during the propagation of electromagnetic waves in a vacuum. In what follows, we consider the particular case of plane waves and their propagation.

1.1.4.7. *Electromagnetic energy density, Poynting identity*

In section 1.1.2, we established the continuity equation [1.51] that reflects the charge conservation principle, which is shown below in its simplified form:

$$\vec{\nabla}\cdot\vec{J}+\frac{\partial\rho}{\partial t}=0 \qquad [1.153]$$

Below, we will aim to find an equation analogous to [1.153] that describes the *conservation of electromagnetic energy*.

To do this, we need to find the equivalents of volume charge density ρ and current density vector $\vec{J}$. Intuitively, the electromagnetic energy density w is equivalent to ρ.

Considering the expressions of electrical energy density w_e [1.129] and magnetic energy density w_m [1.135], the electromagnetic energy density $w = w_e + w_m$ is therefore given by the expression:

$$w=\frac{1}{2}\varepsilon_0E^2+\frac{B^2}{2\mu_0} \qquad [1.154]$$

Now we will attempt to find the equivalent of the current density vector. First, we must express the power transferred by the electromagnetic field to a set of electric charges contained in volume V.

Here we will consider the Maxwell-Ampère equation, which contains the current density vector, and the Maxwell-Faraday equation, which contains the temporal differential of the magnetic field. From Table 1.4, we take:

$$\vec{\nabla} \wedge \vec{B} = \mu_0 \left[\vec{J} + \varepsilon_0 \frac{\partial \vec{E}}{\partial t} \right] = \mu_0 \vec{J} + \mu_0 \varepsilon_0 \frac{\partial \vec{E}}{\partial t} \text{ and } \vec{\nabla} \wedge \vec{E} = -\frac{\partial \vec{B}}{\partial t} \qquad [1.155]$$

To obtain an equation between the electric and magnetic fields and the electromagnetic energy density from [1.155], we multiply the Maxwell-Ampère equation by $\vec{E} / \mu_0$ and the Maxwell-Faraday equation by $\vec{B} / \mu_0$, obtaining:

$$\frac{1}{\mu_0} \vec{E} \cdot \vec{\nabla} \wedge \vec{B} = \vec{J} \cdot \vec{E} + \varepsilon_0 \vec{E} \cdot \frac{\partial \vec{E}}{\partial t} = \vec{J} \cdot \vec{E} + \frac{\partial}{\partial t} \left(\frac{\varepsilon_0 E^2}{2} \right)$$

$$\frac{1}{\mu_0} \vec{B} \cdot \vec{\nabla} \wedge \vec{E} = -\frac{1}{\mu_0} \vec{B} \cdot \frac{\partial \vec{B}}{\partial t} = -\frac{\partial}{\partial t} \left(\frac{B^2}{2\mu_0} \right)$$

If we subtract the second equation from the first equation above, we obtain:

$$\frac{1}{\mu_0} (\vec{E} \cdot \vec{\nabla} \wedge \vec{B} - \vec{B} \cdot \vec{\nabla} \wedge \vec{E}) = \vec{J} \cdot \vec{E} + \frac{\partial}{\partial t} \left(\frac{\varepsilon_0 E^2}{2} + \frac{B^2}{2\mu_0} \right) \qquad [1.156]$$

What remains is for us to transform the first term in the left part of [1.156]. For this, let us take advantage of the vector transformation equation:

$$\vec{\nabla} \cdot (\vec{f} \wedge \vec{g}) = \vec{g} \cdot (\vec{\nabla} \wedge \vec{f}) - \vec{f} \cdot (\vec{\nabla} \wedge \vec{g}) \Rightarrow \vec{\nabla} \cdot (\vec{B} \wedge \vec{E}) = \vec{E} \cdot (\vec{\nabla} \wedge \vec{B}) - \vec{B} \cdot (\vec{\nabla} \wedge \vec{E})$$

Considering this result, equation [1.156] is written as:

$$\frac{1}{\mu_0} \vec{\nabla} \cdot (\vec{B} \wedge \vec{E}) = \vec{J} \cdot \vec{E} + \frac{\partial}{\partial t} \left(\frac{\varepsilon_0 E^2}{2} + \frac{B^2}{2\mu_0} \right)$$

Since $\vec{B} \wedge \vec{E} = -\vec{E} \wedge \vec{B}$, we obtain:

$$\vec{\nabla} \cdot \left(\vec{E} \wedge \frac{\vec{B}}{\mu_0} \right) + \vec{J} \cdot \vec{E} + \frac{\partial}{\partial t} \left(\frac{\varepsilon_0 E^2}{2} + \frac{B^2}{2\mu_0} \right) = 0 \qquad [1.157]$$

Let us consider the case where the current density vector is perpendicular to the electric field. In this case, $\vec{J} \cdot \vec{E} = 0$. Considering [1.154], equation [1.157] is written as follows:

$$\vec{\nabla} \cdot \left(\vec{E} \wedge \frac{\vec{B}}{\mu_0} \right) + \frac{\partial \omega}{\partial t} = 0 \qquad [1.158]$$

Equation [1.158] is analogous to the continuity equation [1.153]. The vector $\vec{E} \wedge \frac{\vec{B}}{\mu_0}$ is analogous to the current density vector $\vec{J}$.

The vector, often denoted $\vec{R}$, is known as the *Poynting vector*, given by the following expression:

$$\vec{R} = \vec{E} \wedge \frac{\vec{B}}{\mu_0} \qquad [1.159]$$

As a result, equation [1.157] is known as the *Poynting identity*.

Before moving onto the physical interpretation of the Poynting identity, we first need to clarify the physical meaning of the scalar product $\vec{J} \cdot \vec{E}$. For this, we consider a discrete distribution of charges contained in a domain with volume τ. Here, n is the volume charge density. Each electric charge in domain τ is subjected to an electromagnetic force [1.86]. The charge $dQ = nqd\tau$ contained in the elementary volume $d\tau$ is subjected to the force:

$$\overrightarrow{dF} = dQ(\vec{E} + \vec{v} \wedge \vec{B}) = nq(\vec{E} + \vec{v} \wedge \vec{B})d\tau \qquad [1.160]$$

The power developed by this force is given by the equation:

$$dP = \overrightarrow{dF} \cdot \vec{v} = \left[nq(\vec{E} + \vec{v} \wedge \vec{B})d\tau \right] \cdot \vec{v}$$

So, since $(\vec{v} \wedge \vec{B}) \cdot \vec{v} = 0$:

$$dP = nq\vec{E} \cdot \vec{v}\, d\tau + nq(\vec{v} \wedge \vec{B}) \cdot \vec{v}\, d\tau = nq\vec{E} \cdot \vec{v}\, d\tau = nq\vec{v} \cdot \vec{E}\, d\tau$$

With the current density vector being $\vec{J} = nq\vec{v}$, the previous expression is written as follows:

$$dP = \vec{J} \cdot \vec{E}\, d\tau$$

In the entire volume including all electric charges, we obtain:

$$P = \iiint_{(\tau)} \vec{J} \cdot \vec{E}\, d\tau \qquad [1.161]$$

The power P is equal to the power transferred from the electromagnetic field to all electric charges contained in volume τ.

Using the work-energy theorem, it becomes:

$$dE_c = \sum dW = Pdt$$

Therefore, considering [1.161]:

$$\frac{dE_c}{dt} = \iiint_{(\tau)} \vec{J} \cdot \vec{E}\, d\tau \qquad [1.162a]$$

Result [1.162a] shows that the product $\vec{J} \cdot \vec{E}$ is equal to the volumetric energy density:

$$du_{em} = \frac{dE_c dt}{d\tau}$$

transferred by the electromagnetic field to the electric charges contained in the elementary volume $d\tau$. E_c is the kinetic energy of the charge carriers present in volume τ.

Now we will move onto the physical interpretation of the Poynting identity [1.157] which we will write below as follows:

$$\vec{\nabla}\cdot\vec{R}+\vec{J}\cdot\vec{E}+\frac{\partial\omega}{\partial t}=0$$

This then gives:

$$\vec{\nabla}\cdot\vec{R}=-\left(\vec{J}\cdot\vec{E}+\frac{\partial\omega}{\partial t}\right)$$

Let us integrate the last equality in the whole volume τ. We obtain:

$$\iiint_{(\tau)}\vec{\nabla}\cdot\vec{R}\,d\tau=-\left[\iiint_{(\tau)}(\vec{J}\cdot\vec{E})d\tau+\iiint_{(\tau)}\left(\frac{\partial\omega}{\partial t}\right)d\tau\right] \quad [1.162b]$$

Here S is the surface surrounding volume τ (Figure 1.23).

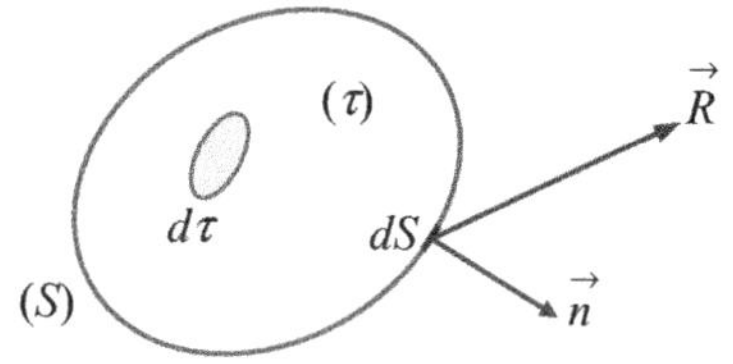

Figure 1.23. *Surface S surrounding volume V containing charge carriers*

Per the divergence theorem:

$$\iiint_{(\tau)}(\vec{\nabla}\cdot\vec{R})d\tau=\iint_{(S)}(\vec{\nabla}\cdot\vec{R})\cdot\vec{dS}=\iint_{(S)}(\vec{\nabla}\cdot\vec{R})\cdot\vec{n}\,dS$$

Using this result and [1.161], equation [1.162b] is written in the form:

$$\iint_{(S)}(\vec{\nabla}\cdot\vec{R})\cdot\vec{n}\,dS=-\left[\frac{dE_c}{dt}+\iiint_{(\tau)}\left(\frac{\partial\omega}{\partial t}\right)d\tau\right] \quad [1.162c]$$

Equation [1.162c] states that the Poynting vector flux over the surface (S) is equal to the decrease in the sum of kinetic energy E_c of the charge carriers and the electromagnetic energy w contained in volume τ.

Also, in contrast to the continuity equation [1.153] expressing the conservation of electric charge, the electromagnetic energy is not conserved.

In the case where the variation in kinetic energy is zero over time (E_c = Cte), equation [1.162c] is written alongside [1.153]:

$$\vec{\nabla} \cdot \vec{R} + \frac{\partial \omega}{\partial t} = 0 \, ; \; \vec{\nabla} \cdot \vec{J} + \frac{\partial \rho}{\partial t} = 0 \qquad [1.163]$$

To express the conservation of electromagnetic energy, we proceed by analogy in the form of a table using equations [1.163]. The results obtained are summarized in Table 1.5.

Conservation of charge	Conservation of electromagnetic energy
ρ: volume charge density	w: electromagnetic energy density
$\vec{J}$: current density vector	$\vec{R}$: Poynting vector
Flux of $\vec{J}$: current intensity equal to the electric charge crossing a surface per unit time	Flux of $\vec{R}$: electromagnetic power equal to the electromagnetic energy crossing a surface per unit time

Table 1.5. *Conservation of charge/conservation of electromagnetic energy analogy*

1.1.4.8. *Vector potential of the progressive electromagnetic plane wave*

Let us consider an axis *Ox* of the unit vector $\vec{e}_x$, and an electromagnetic wave with the propagation direction *Ox*. By definition, a plane wave with direction *Ox*, is a wave whose vector potential is only a function of the *x* coordinate and time *t*: i.e. $\vec{A} = \vec{A}(x,t)$. The components of the vector potential are therefore in the form:

$$A_x\,(x, t);\; A_y\,(x, t);\; A_z\,(x, t)$$

The vector potential satisfies the Coulomb gauge $\vec{\nabla} \cdot \vec{A} = 0$. In Cartesian coordinates, we obtain:

$$\vec{\nabla} \cdot \vec{A} = \frac{\partial A_x(x,t)}{\partial t} + \frac{\partial A_y(x,t)}{\partial t} + \frac{\partial A_z(x,t)}{\partial t} = 0$$

So:

$$\frac{\partial A_x(x,t)}{\partial t} = 0 \Rightarrow A_x(x,t) = A_{0x} = Cte$$

For time-dependent phenomena, choosing a vector potential for component A_{x0} = Cte ≠ 0 is useless. Therefore A_x $(x, t) = A_{0x} = 0$. So, using the propagation equation [1.150] of the vector potential $\vec{A}$ (x, t), the partial differential equations satisfied by components A_y (x, t) and A_z (x, t) are written as follows:

$$\frac{\partial^2 A_y(x,t)}{\partial x^2} = \varepsilon_0\mu_0 \frac{\partial^2 A_y(x,t)}{\partial t^2}; \frac{\partial^2 A_z(x,t)}{\partial x^2} = \varepsilon_0\mu_0 \frac{\partial^2 A_z(x,t)}{\partial t^2} \qquad [1.164]$$

Equations [1.164] express the propagation equations of the vector potential components. As $\varepsilon_0\mu_0 c^2 = 1$, equations [1.150] and [1.152] are written in the general form below:

$$\frac{\partial^2 s(x,t)}{\partial x^2} = \frac{1}{c^2}\frac{\partial^2 s(x,t)}{\partial t^2} \qquad [1.165]$$

To integrate the differential equation [1.165], let $s = s\ (\chi,\mu)$ with the following changes in variables:

$$\chi\,(x, t) = \chi = ct - x;\ \mu\,(x, t) = \mu = ct + x \qquad [1.166]$$

Using [1.149], it becomes:

$$\frac{\partial \chi}{\partial x} = -1;\ \frac{\partial \mu}{\partial x} = +1 \qquad [1.167a]$$

The partial derivatives are expressed as follows:

$$\frac{\partial^2 s(x,t)}{\partial x^2} \text{ and } \frac{\partial^2 s(x,t)}{\partial t^2}$$

We successively obtain [1.167a]:

$$\frac{\partial s}{\partial x} = \frac{\partial s}{\partial \chi}\frac{\partial \chi}{\partial x} + \frac{\partial s}{\partial \mu}\frac{\partial \mu}{\partial x} = -\frac{\partial s}{\partial \chi} + \frac{\partial s}{\partial \mu}$$

$$\frac{\partial^2 s}{\partial x^2}=\frac{\partial}{\partial x}\left(\frac{\partial s}{\partial x}\right)=\frac{\partial}{\partial x}\left(-\frac{\partial s}{\partial \chi}+\frac{\partial s}{\partial \mu}\right)$$

$$\frac{\partial^2 s}{\partial x^2}=-\frac{\partial}{\partial \chi}\left(\frac{\partial \chi}{\partial x}\right)\frac{\partial s}{\partial \chi}-\frac{\partial}{\partial \mu}\left(\frac{\partial \mu}{\partial x}\right)\frac{\partial s}{\partial \chi}+\frac{\partial}{\partial \mu}\left(\frac{\partial \mu}{\partial x}\right)\frac{\partial s}{\partial \mu}+\frac{\partial}{\partial \chi}\left(\frac{\partial \chi}{\partial x}\right)\frac{\partial s}{\partial \mu}$$

So considering [1.167a]:

$$\frac{\partial^2 s}{\partial x^2}=\frac{\partial^2}{\partial \chi^2}-\frac{\partial^2 s}{\partial \mu \partial \chi}+\frac{\partial^2 s}{\partial \mu^2}-\frac{\partial^2 s}{\partial \chi \partial \mu}=\frac{\partial^2}{\partial \chi^2}+\frac{\partial^2 s}{\partial \mu^2}-2\frac{\partial^2 s}{\partial \chi \partial \mu} \qquad [1.167b]$$

Similarly, with the changes to variables [1.166], we obtain:

$$\frac{\partial s}{\partial t}=\frac{\partial s}{\partial \chi}\frac{\partial \chi}{\partial t}+\frac{\partial s}{\partial \mu}\frac{\partial \mu}{\partial t}=c\frac{\partial s}{\partial \chi}+c\frac{\partial s}{\partial \mu}=c\left(\frac{\partial s}{\partial \chi}+\frac{\partial s}{\partial \mu}\right) \qquad [1.167c]$$

$$\frac{\partial^2 s}{\partial t^2}=\frac{\partial}{\partial t}\left(\frac{\partial s}{\partial t}\right)=c\frac{\partial}{\partial t}\left(\frac{\partial s}{\partial \chi}+\frac{\partial s}{\partial \mu}\right)$$

$$\frac{\partial^2 s}{\partial t^2}=c\frac{\partial}{\partial \chi}\left(\frac{\partial \chi}{\partial t}\right)\frac{\partial s}{\partial \chi}+c\frac{\partial}{\partial \mu}\left(\frac{\partial \mu}{\partial t}\right)\frac{\partial s}{\partial \chi}+c\frac{\partial}{\partial \mu}\left(\frac{\partial \mu}{\partial t}\right)\frac{\partial s}{\partial \mu}+c\frac{\partial}{\partial \chi}\left(\frac{\partial \chi}{\partial t}\right)\frac{\partial s}{\partial \mu}$$

So, considering [1.167b]:

$$\frac{\partial^2 s}{\partial t^2}=c^2\frac{\partial^2 s}{\partial \chi^2}+c^2\frac{\partial^2 s}{\partial \mu \partial \chi}+c^2\frac{\partial^2 s}{\partial \mu^2}+c^2\frac{\partial^2 s}{\partial \chi \partial \mu}$$

And after arrangement:

$$\frac{\partial^2 s}{\partial t^2}=c^2\left(\frac{\partial^2 s}{\partial \chi^2}+\frac{\partial^2 s}{\partial \mu^2}+2\frac{\partial^2 s}{\partial \chi \partial \mu}\right) \qquad [1.167d]$$

Using [1.167a] and [1.167c], equation [1.165] gives:

$$\frac{\partial^2}{\partial \chi^2}+\frac{\partial^2 s}{\partial \mu^2}-2\frac{\partial^2 s}{\partial \chi \partial \mu}=\frac{\partial^2 s}{\partial \chi^2}+\frac{\partial^2 s}{\partial \mu^2}+2\frac{\partial^2 s}{\partial \chi \partial \mu}$$

So, after simplification:

$$\frac{\partial^2 s(\chi,\mu)}{\partial\chi\partial\mu}=0 \qquad [1.168]$$

Equation [1.168] can be written in the following form:

$$\begin{cases} \dfrac{\partial}{\partial\mu}\left(\dfrac{\partial s(\chi,\mu)}{\partial\chi}\right)=0 \Rightarrow \dfrac{\partial s(\chi,\mu)}{\partial\chi}=F(\chi) \\ \dfrac{\partial}{\partial\chi}\left(\dfrac{\partial s(\chi,\mu)}{\partial\mu}\right)=0 \Rightarrow \dfrac{\partial s(\chi,\mu)}{\partial\mu}=F(\mu) \end{cases}$$

If $f(\chi)$ and $g(\mu)$ are the primitives of functions $F(\chi)$ and $G(\mu)$, respectively, the general solution to equation [1.168] is in the form:

$$s(\chi,\mu)=f(\chi)+g(\mu)$$

With the changes to variables [1.166], we find:

$$s(x,t)=f(ct-x)+g(ct+x) \qquad [1.169]$$

Physically, the general solution [1.169] satisfies the superposition principle of plane waves. To confirm this assertion, let us place ourselves within the framework of de Broglie's theory of matter waves. In this theory, the plane wave is a matter wave given by the expression:

$$\Psi(\vec{r},t)=\Psi_0 e^{i(\vec{k}\cdot\vec{r}-\omega t)} \Rightarrow \Psi(x,t)=\Psi_0 e^{i(kx-\omega t)}$$

Since the norm of the wave vector is $k=\omega/c$, de Broglie's plane wave, which propagates in the direction of axis Ox and in the direction of the unit vector $\vec{e}_x\left(\vec{k}=\dfrac{\omega}{c}\vec{e}_x\right)$ is written as follows:

$$\Psi(x,t)=\Psi_0 e^{i(x\omega/c-\omega t)}=\Psi_0 e^{-i\omega/c(ct-x)}$$

The plane wave above can be written in the form:

$$\Psi(x,t)=\Psi_0 e^{-ik(ct-x)} \qquad [1.170]$$

So, according to expression [1.170], $f(ct - x)$ is a plane wave that propagates in the direction of axis Ox and in the sense of the unit vector $\vec{e}_x$ (*incident wave*) whereas $g(ct + x)$ is a plane wave that propagates in the direction of axis Ox but in the opposite direction to the unit vector $\vec{e}_x$ (*reflected wave*).

Considering the four scalar functions κ, χ, ψ and ϕ, the components of the vector potential of the plane wave is generally written as follows:

$$\begin{cases} A_x(x,t) = 0 \\ A_y(x,t) = \kappa(ct - x) + \chi(ct + x) \\ A_z(x,t) = \psi(ct - x) + \varphi(ct + x) \end{cases} \quad [1.171]$$

John Henry Poynting was a British physicist. He is known for his work on electromagnetism. He defined the Poynting vector, thus making it possible to account for the principle of energy conservation.

Louis Victor de Broglie was a French mathematician and physicist. He is considered as the founder of wave mechanics thanks to his theory on matter waves or phase waves in 1923. His wave theory led to the advent of quantum mechanics in 1926.

Box 1.11. *Poynting (1852–1914) and de Broglie (1892–1987)*

1.1.4.9. *Transversality of the vector potential and electric and magnetic field of plane waves*

For plane waves, $\vec{\nabla} \cdot \vec{A} = 0$. According to the Lorentz gauge [1.146], the scalar potential satisfies condition $\partial V/\partial t = 0$. In a vacuum, the charge volume density $\rho = 0$. Hence, according to [1.147]:

$$\frac{\partial^2 V}{\partial x^2} = \frac{\partial^2 V}{\partial t^2} = \frac{\partial}{\partial t}\left(\frac{\partial V}{\partial t}\right) = 0 \quad [1.172]$$

Taking [1.165] into account, the temporal variation in the electric field of component E_x is written, according to [1.101]:

$$E_x = -\frac{\partial V}{\partial x} - \frac{\partial A_x(x,t)}{\partial t} = -\frac{\partial V}{\partial x} - \frac{\partial A_{0x}(x,t)}{\partial t} = -\frac{\partial V}{\partial x}$$

With reference to [1.172], we obtain:

$$\frac{\partial E_x}{\partial x} = -\frac{\partial^2 V}{\partial x^2} = 0\,;\ \frac{\partial E_x}{\partial t} = -\frac{\partial}{\partial t}\left(\frac{\partial V}{\partial x}\right) = -\frac{\partial}{\partial x}\left(\frac{\partial V}{\partial t}\right) = 0 \qquad [1.173]$$

The first equality in [1.173] shows that component $E_x(x, t)$ of the electric field is independent of the spatial coordinate x. The second equality shows that $E_x(x, t)$ is also independent of the temporal coordinate t. Consequently component $E_x(x, t) = \text{Cte}$. Therefore, considering a constant component of the electric field for variable phenomena over time is of no use. For plane waves we therefore write:

$$A_x(x, t) = 0;\ E_x(x, t) = 0$$

Now we need to specify the value of component B_x of the magnetic field. Given that $\vec{B} = \vec{\nabla} \wedge \vec{A}$, component B_x is written as follows:

$$B_x = -\frac{\partial A_z(x,t)}{\partial y} - \frac{\partial A_y(x,t)}{\partial z} = 0$$

So, for plane waves, we finally obtain:

$$A_x(x, t) = 0;\ E_x(x, t) = 0;\ B_x(x, t) = 0 \qquad [1.174]$$

Results [1.174] express that the respective components A_x, E_x and B_x of the vector potential, electric and magnetic field of the plane wave are zero at all points in space and time. Consequently, vectors $\vec{A}$, $\vec{E}$ and $\vec{B}$ are perpendicular to the propagation direction Ox. For this reason, the vector potential and the electric and magnetic fields of the plane wave are said to be *transverse*.

1.1.4.10. *Monochromatic progressive plane waves*

By definition, a progressive plane wave, is a plane wave such that one of its scalar functions $f(ct - x)$ and $g(ct + x)$ of solution [1.169] is always zero. Let's put arbitrarily $\chi(ct + x) = \phi(ct + x) = 0$ in [1.171]. The components of the vector potential of the plane progressive are written as follows:

$$A_y(x, t) = \kappa(ct - x);\ A_z(x, t) = \psi(ct - x) \qquad [1.175a]$$

Also, the components of the electric and magnetic fields are associated with the vector potential components of the plane wave by the equations:

$$E_y = -\frac{\partial V}{\partial y} - \frac{\partial A_y(x,t)}{\partial t}; E_z = -\frac{\partial V}{\partial z} - \frac{\partial A_z(x,t)}{\partial t} \qquad [1.175b]$$

$$B_y = -\frac{\partial A_x(x,t)}{\partial z} - \frac{\partial A_z(x,t)}{\partial x}; B_z = -\frac{\partial A_y(x,t)}{\partial x} - \frac{\partial A_x(x,t)}{\partial y} \qquad [1.175c]$$

Since $\partial V/\partial t = 0$, considering [1.175a] we obtain:

$$E_y = -c\kappa'(ct-x); E_z = -c\psi'(ct-x) \qquad [1.175d]$$

$$B_y = \psi'(ct-x); B_z = -\kappa'(ct-x) \qquad [1.175e]$$

In summary, for the progressive electromagnetic plane wave, the expressions of the components of the vector potential and the electric and magnetic fields are written according to equations [1.175]:

$$A_y(x,t) = \kappa(ct-x); E_y = -c\kappa'(ct-x); B_y = \psi'(ct-x) \qquad [1.176]$$

$$A_z(x,t) = \psi(ct-x); E_z = -c\psi'(ct-x); B_z = -\kappa'(ct-x)$$

In equations [1.176], the general functions f' represent the derivative of the scalar function f with respect to time; that is to say: $f' = \partial f/\partial t$.

Equations [1.176] establish two important properties of the electric and magnetic fields of progressive electromagnetic plane waves.

The *1st property* expresses the scalar product $\vec{E}\cdot\vec{B}$. We get the following considering equations [1.174] and [1176]:

$$\vec{E}\cdot\vec{B} = E_yB_y + E_zB_z = -c\kappa'\psi' + c\psi'\kappa' = 0$$

Fields $\vec{E}$ and $\vec{B}$ of the progressive plane wave are therefore *perpendicular*.

The *2nd property* expresses the modulus of fields $\vec{E}$ and $\vec{B}$. Considering equations [1.174] and [1.176] we obtain:

$$E^2 = E_y{}^2 + E_z^2 = c^2\kappa'^2 + c^2\psi'^2 = c^2(\kappa'^2 + \psi'^2)$$

$$B^2 = (\kappa'^2 + \psi'^2) \Rightarrow E^2 = c^2 B^2$$

The ratio of the modulus of fields $\vec{E}$ and $\vec{B}$ of progressive plane waves is constant and equal to the propagation velocity of electromagnetic waves in a vacuum. So: $B = E/c$.

Considering Figure 1.24, we see that $\vec{e}_X$, $\vec{E}$ and $\vec{B}$ form a direct trihedron. So:

$$\vec{B} = \frac{\vec{e}_X}{c} \wedge \vec{E} \qquad [1.177]$$

Electromagnetic waves are generally polychromatic waves. This is the case of light waves radiated by the sun. In some cases, the light wave consists of a single radiation wavelength λ. Such a wave is deemed monochromatic. By definition, a progressive plane wave is monochromatic, if the components of the vector potential are in the form:

$$\begin{cases} A_y(x,t) = A_{0y} \cos\left[\dfrac{2\pi c}{\lambda}\left(t - \dfrac{x}{c}\right) + \phi_{0y}\right] \\ A_z(x,t) = A_{0z} \cos\left[\dfrac{2\pi c}{\lambda}\left(t - \dfrac{x}{c}\right) + \phi_{0z}\right] \end{cases} \qquad [1.178]$$

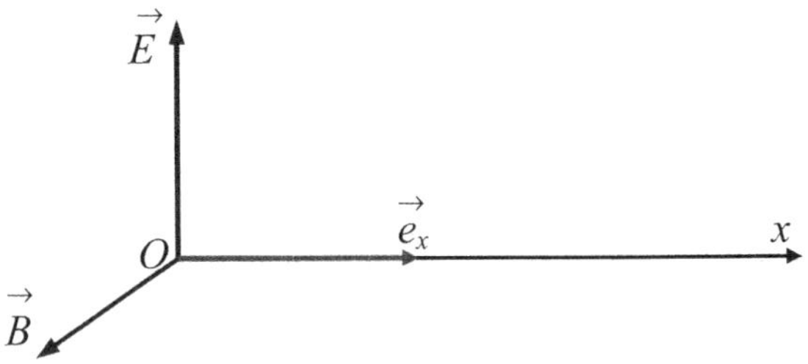

Figure 1.24. *Electric and magnetic fields perpendicular and orthogonal to the propagation direction of the progressive plane wave*

In [1.178], the constants A_{0y} and A_{0z} represent respectively the amplitudes of waves functions $A_y\,(x, t)$ and $A_z\,(x, t)$ and constants φ_{0y} and φ_{0z} are their respective phases.

These amplitudes and phases are the physical variables that characterize the polarization of progressive monochromatic plane waves:

– the polarization is said to be *rectilinear* if $\varphi_{0y} = \varphi_{0z}$: in this case, wave functions $A_y(x, t)$ and $A_z(x, t)$ are "in phase";

– the polarization is said to be *elliptical* if $A_{0y} \neq A_{0z}$ and if the difference in phase defined by $\varphi = \varphi_{0z} - \varphi_{0y}$ is equal to 0 or π. Right or left elliptical polarizations follow the direction of the rotation of the electric field on the ellipse;

– the polarization is said to be *circular* if $A_{0y} = A_{0z}$ and if $\varphi = \varphi_{0z} - \varphi_{0y} = \pm \pi/2$. The polarization is *right circular* if $\varphi = +\pi/2$: the rotation of the electric field on the circle is clockwise (opposite to the trigonometric direction); it is *left circular* if $\varphi = -\pi/2$: the rotation of the electric field on the circle is in the trigonometric direction.

Now we will express the components of the electric and magnetic fields of monochromatic progressive plane waves.

Using [1.75b] as well as components [1.78] of the vector potential, we obtain:

$$\begin{cases} E_y = -\dfrac{\partial A_y(x,t)}{\partial t} = \dfrac{2\pi c}{\lambda} A_{0y} \sin\left[\dfrac{2\pi c}{\lambda}\left(t - \dfrac{x}{c}\right) + \phi_{0y}\right] \\ E_z = -\dfrac{\partial A_z(x,t)}{\partial t} = \dfrac{2\pi c}{\lambda} A_{0z} \sin\left[\dfrac{2\pi c}{\lambda}\left(t - \dfrac{x}{c}\right) + \phi_{0z}\right] \end{cases} \quad [1.179a]$$

$$\begin{cases} B_y = -\dfrac{\partial A_z(x,t)}{\partial x} = -\dfrac{2\pi}{\lambda} A_{0z} \sin\left[\dfrac{2\pi c}{\lambda}\left(t - \dfrac{x}{c}\right) + \phi_{0z}\right] \\ B_z = \dfrac{\partial A_y(x,t)}{\partial x} = \dfrac{2\pi}{\lambda} A_{0y} \sin\left[\dfrac{2\pi c}{\lambda}\left(t - \dfrac{x}{c}\right) + \phi_{0y}\right] \end{cases} \quad [1.179b]$$

Let us consider the particular case where $\varphi_{0y} = \varphi_{0z} = 0$. Under these conditions, the vector expressions of the vector potential $\vec{A}$ and fields $\vec{E}$ and $\vec{B}$ are written as:

$$\begin{cases} \vec{A}(x,t) = A_{0y} \cos\dfrac{2\pi c}{\lambda}\left(t - \dfrac{x}{c}\right)\vec{e}_x + A_{0z} \cos\dfrac{2\pi c}{\lambda}\left(t - \dfrac{x}{c}\right)\vec{e}_z \\ \vec{A}(x,t) = \vec{A}_0 \cos\dfrac{2\pi c}{\lambda}\left(t - \dfrac{x}{c}\right); \; \vec{A}_0 = A_{0y}\vec{e}_x + A_{0z}\vec{e}_z \end{cases} \quad [1.180a]$$

$$\begin{cases} \vec{E}(x,t) = \dfrac{2\pi c}{\lambda} A_{0y} \sin \dfrac{2\pi c}{\lambda}\left(t - \dfrac{x}{c}\right)\vec{e}_y + \dfrac{2\pi c}{\lambda} A_{0z} \sin \dfrac{2\pi c}{\lambda}\left(t - \dfrac{x}{c}\right)\vec{e}_z \\ \vec{E}(x,t) = \vec{E}_0 \sin \dfrac{2\pi c}{\lambda}\left(t - \dfrac{x}{c}\right) \\ \vec{E}_0 = \dfrac{2\pi c}{\lambda}\left(A_{0y}\vec{e}_x + A_{0z}\vec{e}_z\right) = \dfrac{2\pi c}{\lambda}\vec{A}_0 \end{cases} \qquad [1.180b]$$

$$\begin{cases} \vec{B}(x,t) = -\dfrac{2\pi}{\lambda} A_{0z} \sin \dfrac{2\pi c}{\lambda}\left(t - \dfrac{x}{c}\right)\vec{e}_y + \dfrac{2\pi}{\lambda} A_{0y} \sin \dfrac{2\pi c}{\lambda}\left(t - \dfrac{x}{c}\right)\vec{e}_z \\ \vec{B}(x,t) = \vec{B}_0 \sin \dfrac{2\pi c}{\lambda}\left(t - \dfrac{x}{c}\right) \\ \vec{B}_0 = \dfrac{2\pi}{\lambda}\left(-A_{0z}\vec{e}_y + A_{0y}\vec{e}_z\right) = \dfrac{2\pi}{\lambda}\vec{e}_x \wedge \vec{A}_0 \end{cases} \qquad [1.180c]$$

Fields $\vec{E}$ and $\vec{B}$ of monochromatic progressive plane waves verify the properties of progressive plane waves, which are listed below:

– $\vec{E}$ and $\vec{B}$ are perpendicular at all points in space and time;

– the equation linking the modulus of fields $\vec{E}$ and $\vec{B}$ is c: $B = E/c$;

– the vectors $\vec{e}_x$, $\vec{E}$ and $\vec{B}$ form a direct trihedron and verify equation [1.177].

In addition to the properties of progressive plane waves, fields $\vec{E}$ and $\vec{B}$ are characterized by special properties when the progressive plane wave is monochromatic. Using [1.180], we see that for a monochromatic progressive plane wave:

– the vector potential $\vec{A}$ has a constant direction equal to that of $\vec{A}_0$;

– the field $\vec{E}$ has a constant direction which is that of its amplitude:

$$\vec{E}_0 = (2\pi c / \lambda)\vec{A}_0$$

– the field $\vec{B}$ has a constant direction which is that of its amplitude:

$$\vec{B}_0 = (2\pi / \lambda)^2\, \vec{e}_x \wedge \vec{A}_0$$

– the field $\vec{E}$ is collinear to the vector potential $\vec{A}$ at all points in space and time.

1.1.4.11. *Maxwell's equations in a perfect metal*

Metals are characterized by their specific properties to conduct electric current. When a metal is subjected to an electric field $\vec{E}$, electrons mobilize with a velocity $\vec{v}$. If μ is the mobility of electrons and ρ_m is the charge volume density of mobile electrons, the velocity $\vec{v}$ and the current density vector $\vec{J}$ are written as follows:

$$\vec{v} = \mu\vec{E}\,;\ \vec{J} = \rho_m\vec{v} = \rho_m\mu\vec{E} = \sigma\vec{E} \qquad [1.181]$$

By definition, the physical variable $\sigma = \rho_m\mu$ is known as the metal conductivity.

A metal is said to be *perfect* if its conductibility is infinite. Since the charge volume density and the electron conduction velocity are finite, then for a perfect metal, the electric field *is uniformly zero*. According to the Maxwell-Gauss equation:

$$\vec{\nabla}\cdot\vec{E} = \frac{\rho_m}{\varepsilon_0} = 0 \Rightarrow \rho_m = 0$$

In conclusion, in a perfect metal, the charge volume density ρ_m is uniformly zero.

In addition, Maxwell's equations in a vacuum are valid in a perfect metal. Using the Maxwell-Faraday equation, it becomes:

$$\vec{\nabla}\wedge\vec{E} = -\frac{\partial\vec{B}}{\partial t} = \vec{0} \Rightarrow \vec{B} = \overrightarrow{Cte}$$

The result above shows that the magnetic field can only be constant in a perfect metal. However, for time-dependent phenomena, considering a uniform magnetic field is of no use. So, in a perfect metal, the magnetic field is also uniformly zero. Since $\vec{B} = \vec{\nabla}\wedge\vec{A}$, the vector potential $\vec{A}$ is also *uniformly zero* in a perfect metal. Finally, using the Maxwell-Ampère equation, we find that the current density vector is also *uniformly zero* in a perfect metal:

$$\vec{\nabla}\wedge\frac{\vec{B}}{\mu_0} = \vec{J} + \varepsilon_0\frac{\partial\vec{E}}{\partial t},\ \vec{B} = \vec{0},\ \vec{E} = \vec{0} \Rightarrow \vec{J} = \vec{0}$$

So, in summary, in a perfect metal:

$$\vec{E} = \vec{0}\,;\ \vec{B} = \vec{0}\,;\ \vec{A} = \vec{0}\,;\ \rho = 0;\ \vec{J} = \vec{0} \qquad [1.182]$$

1.1.4.12. *Reflection of the electromagnetic field on the surface of a perfect metal*

Here we consider a monochromatic electromagnetic progressive plane wave propagating in a vacuum. Ox is the unit vector propagation axis $\vec{e}_x$, ω is the pulsation and $\vec{k} = \frac{\omega}{c}\vec{e}_x$ is the wave vector of the progressive plane wave.

During propagation, the wave meets the flat surface of a perfect metal arranged vertically as shown in Figure 1.25. In what follows, we focus on the reflection process of the plane wave on the metal occupying the half-space $x \geq 0$ (the thickness of the metal does not come into play in this study of the reflection process; therefore it can be disregarded).

The electromagnetic field close to the metal surface is the superposition of an incident wave and a reflected wave. Here $\vec{E}_i$ and $\vec{E}_r$ are components of the electric field of the incident wave and the reflected wave respectively. Under the principle of superposition, the resulting electric field is equal to:

$$\vec{E} = \vec{E}_i + \vec{E}_r$$

$\vec{B}_i$ and $\vec{B}_r$ are the components of the magnetic field of the incident wave and reflected wave, respectively:

$$\vec{B} = \vec{B}_i + \vec{B}_r$$

Let us assume that the monochromatic progressive plane wave is rectilinearly polarized according to axis Oz.

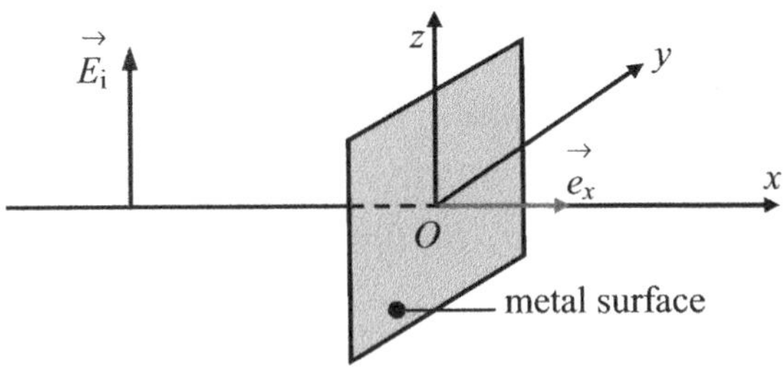

Figure 1.25. *Perfect metal arranged vertically on the propagation axis of a monochromatic progressive plane wave*

Let us consider s (x, t) as one of the components of vectors $\vec{A}$, $\vec{E}$ or $\vec{B}$. In complex notation, the incident $s_i^*(x,t)$ and reflected $s_r^*(x,t)$ plane waves are written:

$$\begin{cases} s_i^*(x,t) = s_{0i}\, e^{i(k_i x - \omega t)} \\ s_r^*(x,t) = s_{0r}\, e^{i(k_r x - \omega t)} \end{cases} \qquad [1.183]$$

In equations [1.183] we write $\omega_i = \omega_r = \omega$. Indeed, the incident wave and the reflected wave both propagate in a vacuum. They therefore have the same pulsation equal to ω.

Before reflection, the electric field of the rectilinearly polarized plane wave according to Oz is combined with its component $\vec{E}_i$ (Figure 1.25). For example:

$$E_{ix} = 0\,;\, E_{iy} = 0\,;\, E_{iz} = E_{0i}\, e^{i(k_i x - \omega t)} \;\; ; \;\; k_i = k = \frac{\omega}{c} \qquad [1.184]$$

After reflection, the components of the electric field $\vec{E}_r$ are written according to the general equations [1.183]:

$$E_{rx} = 0\,;\, E_{ry} = 0\,;\, E_{rz} = E_{0r}\, e^{i(k_r x - \omega t)} \qquad [1.185]$$

Next, we apply the continuity conditions imposed on all wave functions in contact with a surface of separation to explain the expressions of components E_{rx}, E_{ry}, E_{rz} of the electric field and component k_r of the wave vector of the reflected wave.

Knowing that in a perfect metal, the electric field E_m is uniformly zero [1.182], the continuity of the electric field at point O implies that: $E = E_i + E_r = E_m$. Therefore:

$$E_{ix} + E_{rx} = E_{mx} = 0 \Rightarrow E_{ix} = -\,E_{rx} = 0$$

$$E_{iy} + E_{ry} = E_{my} = 0 \Rightarrow E_{iy} = -\,E_{ry} = 0$$

$$E_{iz} + E_{rz} = E_{mz} = 0 \Rightarrow E_{iz} = -\,E_{rz} \Rightarrow E_{0i}\, e^{i(k_i x - \omega t)} = -E_{0r}\, e^{i(k_r x - \omega t)}$$

Choosing the initial conditions so that the amplitude s_0 of the plane wave [1.183] is real, we obtain whilst considering the latter of the equalities above since s_0 is positive:

$$(E_{iz})^2 = (E_{rz})^2 \Rightarrow |E_{0i}|^2 = |E_{0r}|^2 = |E_0|^2 \Rightarrow E_{0i} = E_{0r} = E_0 \qquad [1.186]$$

Moreover, the last of the above equations is true regardless of the time t and especially for $t = 0$. Since $E_{0i} = E_{0r} = E_0$:

$$E_{0i}\, e^{ik_i x} = -E_{0r}\, e^{ik_r x} \Rightarrow \cos k_i x + i \sin k_i x = -\cos k_r x - i \sin k_r x$$

The last of the above equations is verified if cos $(k_i x) = -$ cos $(k_r x)$ and sin $(k_i x) = -$ sin $(k_i x)$. With the functions cosine and sine being even and odd, respectively, the equation between the wave numbers k_i and k_r can be determined using the equation sin $(k_i x) = -$ sin $(k_r x) =$ sin $(-k_r x)$. Whereby, $k_i x = -k_r x$.

Here, the norm of the wave vector of the reflected wave is such that:

$$k_i = -k_r = \frac{\omega}{c} \Rightarrow k_r = -k = -\frac{\omega}{c} \qquad [1.187]$$

The components of the reflected electric field [1.185] are written as follows:

$$E_{rx} = 0\,;\, E_{ry} = 0\,;\, E_{rz} = -E_0\, e^{i(-kx-\omega t)}\,;\, k = \frac{\omega}{c} \qquad [1.188]$$

In summary, the incident and reflected electric fields are written:

$$\begin{cases} \vec{E}_i = E_0\, e^{i(kx-\omega t)}\, \vec{e}_x \\ \vec{E}_r = -E_0\, e^{i(-kx-\omega t)}\, \vec{e}_x \\ k = \dfrac{\omega}{c} \end{cases} \qquad [1.189]$$

Expression [1.189] shows that the reflected electric field $\vec{E}_r$ corresponds to a monochromatic progressive plane wave propagating in the negative sense.

What is left to study is the behavior of the magnetic field in contact with the monochromatic plane wave and the perfect metal. For this, we use result [1.177] and then express the equations between the incident, electric and magnetic fields and between the reflected electric and magnetic fields as follows:

$$\vec{B}_i = \frac{\vec{e}_x}{c} \wedge \vec{E}_i \; ; \; \vec{B}_r = \frac{\vec{e}_x}{c} \wedge \vec{E}_r \qquad [1.190]$$

Considering [1.189], the components $\vec{B}_i$ and $\vec{B}_r$ are written as follows:

$$\begin{cases} B_{ix} = 0 \; ; \; B_{iz} = 0 \; ; \; B_{iy} = \dfrac{E_0}{c} e^{i(kx-\omega t)} \\ B_{rx} = 0 \; ; \; B_{rz} = 0 \; ; \; B_{ry} = -\dfrac{E_0}{c} e^{i(-kx-\omega t)} \end{cases} \qquad [1.191]$$

In the vector form, the incident and reflected magnetic fields are written as:

$$\begin{cases} \vec{B}_i = B_0 \, e^{i(kx-\omega t)} \vec{e}_y \\ \vec{B}_r = - B_0 \, e^{i(-kx-\omega t)} \vec{e}_y \\ B_0 = \dfrac{E_0}{c} ; k = \dfrac{\omega}{c} \end{cases} \qquad [1.192]$$

1.1.4.13. *Superposition of incident and reflected waves: stationary wave*

Let us express the fields $\vec{E} = \vec{E}_i + \vec{E}_r$ and $\vec{B} = \vec{B}_i + \vec{B}_r$ using [1.189] and [1.192]. We obtain:

$$\begin{cases} E_x = E_y = 0 \, ; E_z = E_0 \, e^{i(kx-\omega t)} - E_0 \, e^{i(-kx-\omega t)} = E_0 \, e^{-i\omega t} (e^{ikx} - e^{-ikx}) \\ B_x = B_z = 0 \, ; B_y = - B_0 \, e^{i(kx-\omega t)} - B_0 \, e^{i(-kx-\omega t)} = -B_0 \, e^{-i\omega t} (e^{ikx} + e^{-ikx}) \end{cases}$$

Using trigonometric functions, we obtain:

$$E_z = E_0 \, e^{-i\omega t} [\cos kx + i \sin kx - (\cos kx - i \sin kx)] = 2iE_0 \, e^{-i\omega t} \sin kx$$

$$B_y = -B_0 \, e^{-i\omega t} [\cos kx + i \sin kx + (\cos kx - i \sin kx)] = -2B_0 \, e^{-i\omega t} \cos kx$$

So:

$$E_z = 2iE_0 (\cos \omega t - i \sin \omega t) \sin kx = 2iE_0 \cos \omega t \sin kx + 2E_0 \sin \omega t \sin kx$$

$$B_y = -2B_0 (\cos \omega t - i \sin \omega t) \cos kx = -2B_0 \cos \omega t \cos kx + 2iB_0 \sin \omega t \cos kx$$

In real notation, the expression of the plane wave $s\ (x, t)$ is equal to the real part of the wave function $s^*\ (x, t)$ [1.183] in complex notation. So $s = \Re(s^*)$ where $\Re$ is the real part of the complex wave s^*. Using real notation, which is more suitable for the discussions that follow, we obtain:

$$\begin{cases} E_z = 2E_0 \sin \omega t \sin kx \\ B_y = -2B_0 \cos \omega t \cos kx \\ B_0 = E_0 / c\, ; k = \omega / c \end{cases} \qquad [1.193]$$

By definition, the stationary wave is any wave resulting from the superposition of two waves of the same pulsation and amplitude, but which propagate in opposite directions. The wave numbers have the same value equal to k for both waves but with opposite signs. Such a wave does not propagate, instead it produces an oscillation in place, hence the term "stationary". Mathematically, a stationary wave with propagation direction Ou is the product of a function of space coordinates by a function of time with the form:

$$s\ (u, t) = s_0 \cos ku \cos \omega t \text{ or } s\ (u, t) = s_0 \sin ku \sin \omega t \qquad [1.194]$$

Components E_z and B_y [1.193] of the electric field and magnetic field respectively are characterized by a stationary wave. It is no longer progressive since there is no propagation. As with all stationary waves, the vibration of fields E_z and B_y is characterized by the existence of *nodes* and *anti-nodes*. By definition:

– the *nodes* are the vibration minima of the stationary wave at points u fixed in space;

– the *anti-nodes* are the vibration maxima of the stationary wave at points u fixed in space.

Moreover, there are *nodal planes* for which the stationary wave is zero at all instants. Let us specify these planes for the total electric and magnetic fields [1.194] of the stationary electromagnetic wave.

The amplitudes of the electric and magnetic fields depend on variable x.

– For the electric field, the nodal planes are situated at points x such that:

$$\sin kx = 0 \Rightarrow kx_n = n\pi\ ; k = \frac{\omega}{c} = \frac{2\pi}{\lambda} \Rightarrow x_n = \frac{n\lambda}{2}\ ; n \text{ is an integer}$$

– For the magnetic field, the nodal planes are situated at points x such that:

$$\cos kx = 0 \Rightarrow kx_{\ p} = \frac{(2p+1)\pi}{2}\ ; k = \frac{2\pi}{\lambda} \Rightarrow x_p = \frac{p\lambda}{2} + \frac{\lambda}{4}\ ; p \text{ is an integer}$$

The nodal planes of the magnetic field are offset by $\lambda/4$ with respect to the nodal planes of the electric field. With the metal plane situated at $x = 0$ (Figure 1.26), the metal surface is a nodal plane for the electric field. A node corresponds to point O for the electric field and an antinode for the magnetic field, as indicated in Figure 1.26, illustrating the structure of the electromagnetic stationary wave.

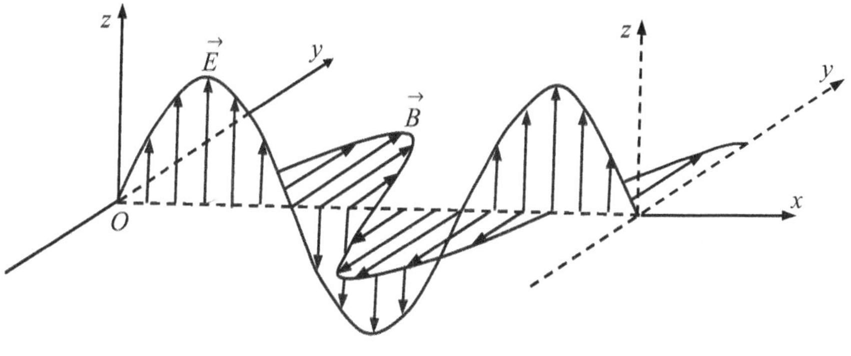

Figure 1.26. *Structure of an electromagnetic stationary wave*

1.2. Maxwell equations in material media[4]

We previously studied electromagnetics in a vacuum made up of the basic disciplines of electrostatics, magnetostatics and electromagnetic induction. Then, we deduced from this study, the fundamental equations of electromagnetism comprising Maxwell's equations in a vacuum of charges and currents.

4 (Annequin and Boutigny 1974; Bok and Hulin-Jung 1979; Faroux and Renault 1998; Mauras 1999; Stöcker *et al.* 2007).

The objective of this section is to study electromagnetism in material media with the aim to establish Maxwell's equations in these media. On a microscopic scale, matter is made up of atoms that can unite to form molecules. Each atom consists of a nucleus and an electron cloud. The charged particles of the matter consist of electrons and protons interacting strongly with electric fields. In addition, these particles are in motion and are characterized by their magnetic spin and orbital moments. Matter is thus the site of polarization and magnetization phenomena.

In this section, we will therefore introduce the notions of polarization vector $\vec{P}$ and magnetization vector $\vec{M}$. To establish Maxwell's equations in a material medium, two new vector fields will be introduced: the electric induction vector $\vec{D}$ (or even the electric displacement vector) and the magnetic excitation vector $\vec{H}$. Maxwell's equations in a material medium therefore involve four vector fields: the electric field $\vec{E}$, the magnetic field $\vec{B}$ and the fields $\vec{D}$ and $\vec{H}$. We will limit our study to linear, homogeneous and isotropic dielectric media.

1.2.1. *Electric field and potential in macroscopic dielectric media*

1.2.1.1. *Dielectric medium, polarization vector*

We clarified in the introduction that charged particles of matter are characterized by their magnetic spin and orbital moments. The action of an external electric field directs the elementary electric dipoles in well-defined directions. For an electric dipole with a dipole moment $\vec{p}$ placed in an electric field $\vec{E}$, the interaction energy potential between the dipole and electric field is defined by the following equation:

$$W = -\vec{p} \cdot \vec{E} \qquad [1.195]$$

For example, regarding polar molecules, the most likely arrangement of the dipole moments of the most stable molecules corresponds to the attraction of the dipoles, which decreases the potential energy [1.195]. The dipole moments are then oriented parallel and antiparallel to the electric field (Figure 1.27). On a macroscopic scale, the permanent polarization of matter or that induced by the action of an external electric field, causes the creation of a polarization vector $\vec{P}$. In a space domain with volume $d\tau$ of polarized matter, the elementary electric dipole moment $d\vec{M}$ is defined by the equation:

$$\overrightarrow{dM} = \vec{P}\, d\tau \qquad [1.196]$$

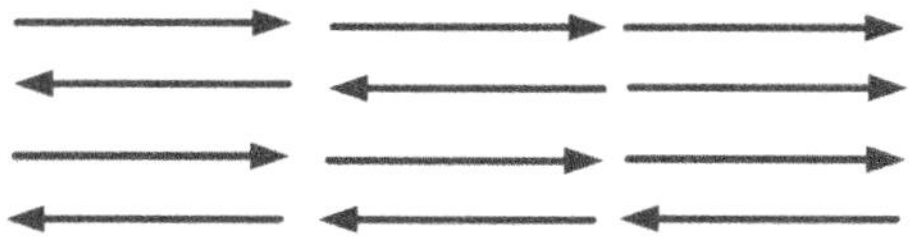

Figure 1.27. *Most likely arrangement of dipole moments of the most stable polar molecules*

The polarization vector can be both uniform and not in volume $d\tau$ and can depend on time. In what follows, we will assume it is independent of time.

1.2.1.2. *Scalar potential in a polarized environment, polarization charges*

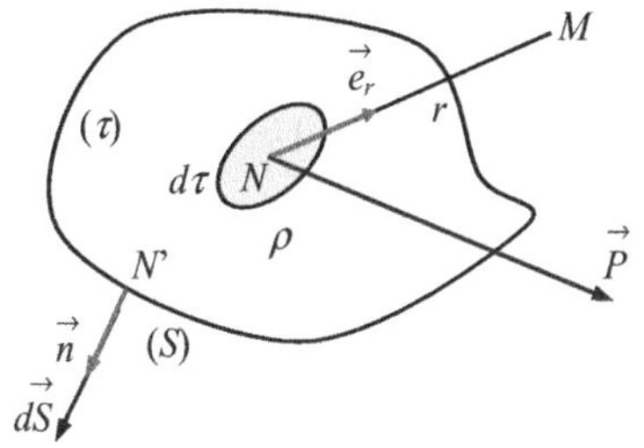

Figure 1.28. *Dielectric medium with volume τ*

Let us consider a dielectric medium occupying a certain space domain with volume τ (Figure 1.28). This domain is the site of permanent polarization or is induced and characterized by a free-charge volume distribution. At point N in this domain, the polarization vector is $\vec{P}$ and the free-charge (ρ) volume density is denoted ρ_{free}.

The electric potentials dV_ρ and dV_p respectively created in M by the charge volume distribution and by the volume distribution of electric dipoles characterized by the polarization vector $\vec{P}$ are deduced from the electric scalar potential [1.70a] and the electric dipole potential [1.80] respectively:

$$dV_\rho = \frac{dq}{4\pi\varepsilon_0\, r} = \frac{\rho_{free}\, d\tau}{4\pi\varepsilon_0\, r};\; dV_p = \frac{\overrightarrow{dM}\cdot\vec{e}_r}{4\pi\varepsilon_0\, r^2} = \frac{\vec{P}\cdot\vec{e}_r\, d\tau}{4\pi\varepsilon_0\, r^2} \qquad [1.197]$$

The overall scalar potential at point M is therefore given by the superposition principle, i.e. $dV = dV_\rho$ and dV_p. For all free charges and electric dipoles distributed in volume τ, we obtain:

$$V = \iiint_{(\tau)} \frac{\rho_{free}\, d\tau}{4\pi\varepsilon_0\, r} + \iiint_{(\tau)} \frac{\vec{P}\cdot\vec{e}_r\, d\tau}{4\pi\varepsilon_0\, r^2} \qquad [1.198]$$

Besides, due to the polarization, charges distributed in the volume (τ) and others distributed on the surface (S) of the domain considered appear in the medium (Figure 1.28). These charges are called polarization charges and are responsible for the scalar potential dV_ρ created by the polarization of the medium. We seek to explain this in what follows.

Note first of all that the overall potential [1.198] contains a first term in $1/r$ and a second term in $1/r^2$. To introduce the polarization charges, the second term is transformed so as to obtain a potential in $1/r$. For this, we consider the gradient of the scalar function $f(r)$:

$$\vec{\nabla} f = \frac{\partial f}{\partial r}\vec{e}_r + \frac{1}{r}\frac{\partial f}{\partial \theta}\vec{e}_\theta + \frac{1}{r\sin\theta}\frac{\partial f}{\partial \varphi}\vec{e}_\varphi$$

For the radial function $f(r) = 1/r$, we get:

$$\vec{\nabla}\left(\frac{1}{r}\right) = \frac{d}{dr}\left(\frac{1}{r}\right)\vec{e}_r = \frac{\vec{e}_r}{r^2} \qquad [1.199a]$$

Now we make use of the following vector transformation equations:

$$\vec{\nabla}\left(\frac{1}{r}\right) = \frac{\vec{e}_r}{r^2}\vec{\nabla}\cdot\vec{P} + \vec{P}\cdot\vec{\nabla}\left(\frac{1}{r}\right);\ \vec{\nabla}\cdot\left(\frac{\vec{P}}{r}\right) = \frac{1}{r}\vec{\nabla}\cdot\vec{P} + \vec{P}\cdot\vec{\nabla}\left(\frac{1}{r}\right) \qquad [1.199b]$$

Using [1.199a], the final equality [1.199b] gives:

$$\vec{\nabla}\cdot\left(\frac{\vec{P}}{r}\right) = \frac{\vec{\nabla}\cdot\vec{P}}{r} + \frac{\vec{P}\cdot\vec{e}_r}{r^2} \Rightarrow \frac{\vec{P}\cdot\vec{e}_r}{r^2} = \vec{\nabla}\cdot\left(\frac{\vec{P}}{r}\right) - \frac{\vec{\nabla}\cdot\vec{P}}{r} \qquad [1.199c]$$

Inserting the second equality [1.199c] into [1.198], we find:

$$V = \iiint_{(\tau)} \frac{\rho_{free}}{4\pi\varepsilon_0} \frac{d\tau}{r} + \iiint_{(\tau)} \frac{1}{4\pi\varepsilon_0} \vec{\nabla} \cdot \left(\frac{\vec{P}}{r} \right) d\tau - \iiint_{(\tau)} \frac{1}{4\pi\varepsilon_0} \frac{\vec{\nabla} \cdot \vec{P}}{r} d\tau \quad [1.199d]$$

The last term to the left of the expression [1.199d] is analogous to the expression of the scalar potential created by a charge volume distribution characterized by the charge volume density $\vec{\nabla} \cdot \vec{P}$. So, $\rho_{pol} = -\vec{\nabla} \cdot \vec{P}$. The expression [1.199d] is written as follows:

$$V = \iiint_{(\tau)} \frac{\rho_{free}}{4\pi\varepsilon_0} \frac{d\tau}{r} + \iiint_{(\tau)} \frac{1}{4\pi\varepsilon_0} \vec{\nabla} \cdot \left(\frac{\vec{P}}{r} \right) d\tau + \iiint_{(\tau)} \frac{\rho_{pol}}{4\pi\varepsilon_0} \frac{d\tau}{r} d\tau \quad [1.199e]$$

Now we will transform the second term in the right-hand side of the expression [1.199e].

Let *N'* be a point on the closed surface (*S*) surrounding the volume domain (τ) (Figure 1.28). The polarization vector at *N'* is written $\vec{P}' = \vec{P}(N)$. Using the divergence theorem we obtain:

$$\iiint_{(\tau)} \vec{\nabla} \cdot \left(\frac{\vec{P}}{r} \right) d\tau = \iint_{(S)} \frac{\vec{P}' \cdot \overrightarrow{dS}}{r} = \iint_{(S)} \frac{\vec{P}' \cdot \vec{n}\, dS}{r} = \iint_{(S)} \frac{\sigma_{pol}\, dS}{r} \quad [1.199f]$$

In equation [1.199f], the surface charge density of the polarization charges distributed over surface (*S*) is $\sigma_{pol} = \vec{P}' \cdot \vec{n} = \vec{P}(N') \cdot \vec{n}$. Using the last term to the left of the expression [1.199f], the overall potential [1.199e] at point *M* created by all the free charges and electric dipoles distributed in volume (τ) is written as follows:

$$V = \iiint_{(\tau)} \frac{\rho_{pol}\, d\tau}{4\pi\varepsilon_0 r} + \iiint_{(\tau)} \frac{\sigma_{pol}\, dS}{4\pi\varepsilon_0 r} + \iiint_{(\tau)} \frac{\rho_{pol}\, d\tau}{4\pi\varepsilon_0 r} d\tau \quad [1.200]$$

By definition, the polarization charges are:

– the charges due to the polarization of the medium and distributed over the surface (*S*). They are characterized by the surface charge density $\sigma_{pol} = \vec{P}' \cdot \vec{n} = \vec{P}(N') \cdot \vec{n}$;

– the charges due to the polarization of the medium and distributed in the volume (τ). They are characterized by the volume charge density $\rho_{pol} = -\vec{\nabla} \cdot \vec{P} = -\vec{\nabla} \cdot \vec{P}(N)$;

– the polarization charges are associated with a polarization density vector denoted $\vec{J}_{pol} = \dfrac{\partial \vec{P}}{\partial t}$.

Thus, when the medium has a permanent polarization or is induced by an external electric field, the surface and volume polarization charge densities as well as the polarization density vector are given by the respective equations:

$$\sigma_{pol} = \vec{P} \cdot \vec{n}; \quad \rho_{pol} = -\vec{\nabla} \cdot \vec{P}; \quad \vec{J}_{pol} = \frac{\partial \vec{P}}{\partial t} \qquad [1.201a]$$

The total volume density ρ_{tot} and the total current density vector $\vec{J}_{tot}$ are given by the equations:

$$\rho_{tot} = \rho_{free} + \rho_{pol}; \quad \vec{J}_{tot} = \vec{J}_{free} + \vec{J}_{pol} \qquad [1.201b]$$

When the polarization disappears ($\sigma_{pol} = \rho_{pol} = 0$), only the free charges with charge volume density ρ remain in the medium. The corresponding potential is given by the first term in the general expression [1.200]. Moreover, in a steady state, the polarization vector $\vec{P}$ is constant, whereby $\vec{J}_{pol} = \vec{0}$. Also, the charge volume density of polarization and the volume density of the polarization current verify the local charge conservation equation [1.153]:

$$\vec{\nabla} \cdot \vec{J}_{pol} + \frac{\partial \rho_{pol}}{\partial t} = 0 \qquad [1.202]$$

1.2.1.3. *Electric field and displacement*

In a dielectric medium with a free charge ρ_{free} volume density and polarization charge ρ_{pol} volume density, the local formulation of Gauss's theorem [1.1.25] is written as follows:

$$\vec{\nabla} \cdot \vec{E} = \frac{\rho_{free} + \rho_{pol}}{\varepsilon_0} \qquad [1.203]$$

Since $\rho_{pol} = -\vec{\nabla}\cdot\vec{P}$, equation [1.203] is written in the form:

$$\vec{\nabla}\cdot\vec{E} = \frac{\rho_{free} - \vec{\nabla}\cdot\vec{P}}{\varepsilon_0} \Rightarrow \left[\varepsilon_0 \vec{\nabla}\cdot\vec{E} + \vec{\nabla}\cdot\vec{P}\right] = \vec{\nabla}\cdot\left[\varepsilon_0 \vec{E} + \vec{P}\right] = \rho_{free} \qquad [1.204]$$

The last equality is analogous to equation $\vec{\nabla}\cdot\vec{P} = -\rho_{pol}$. Hence, we define a displacement electric vector $\vec{D}$ given by the equation:

$$\vec{D} = \varepsilon_0 \vec{E} + \vec{P} \qquad [1.205]$$

According to [1.204]:

$$\vec{\nabla}\cdot\vec{D} = \rho_{free} \qquad [1.206]$$

The electric displacement vector $\vec{D}$ is also known as the electric induction vector or even the electric excitation vector. In a vacuum of free charges ($\rho_{\text{free}} = 0$), the divergence of the displacement vector is zero according to [1.206]: its flux is conservative in virtue of the divergence theorem.

1.2.1.4. *Electric field and electric displacement vector continuity equations across the vacuum-dielectric surface of separation*

Here we study the refraction of the electric field and the electric displacement vector across a vacuum-dielectric surface of separation (Figure 1.29).

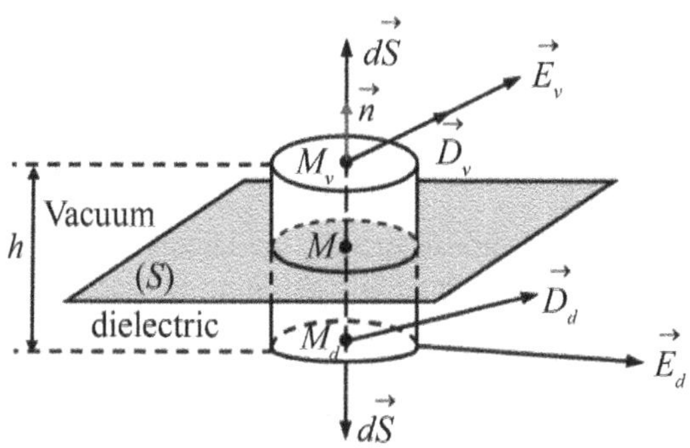

Figure 1.29. *Refraction of the electric displacement vector across a vacuum-dielectric surface of separation*

The electric field and the displacement vector in a vacuum and in the dielectric are respectively: $\vec{E}_v$ and $\vec{D}_v$ and $\vec{E}_d$ and $\vec{D}_d$. So, according to [1.205]:

$$\vec{D}_v = \varepsilon_0 \vec{E}_v \quad ; \quad \vec{D}_d = \varepsilon_0 \vec{E}_d + \vec{P} \qquad [1.207]$$

In the vacuum, the displacement vector $\vec{E}_v$ is proportional to the electric field $\vec{E}_v$ [1.207]. Since $\varepsilon_0 < 1$, then the vectors $\vec{E}_v$ and $\vec{D}_v$ have the same sense and direction (Figure 1.29). However, in the dielectric, the field $\vec{E}_d$ and the displacement vector $\vec{D}_d$ do not have the same direction due to the polarization vector $\vec{P}$.

Let us choose a small cylinder with a base surface area dS and infinitesimal height h (Figure 1.29). We express the total flux of the displacement vector leaving the cylinder as equal to the sum of the fluxes leaving the circular dS and central base surface area M_v and M_d and the flux leaving the lateral surface area of the cylinder. We obtain:

$$\Phi_{tot} = \iint_{(S)} \vec{D}_d \cdot \vec{dS} + \iint_{(S)} \vec{D}_v \cdot \vec{dS} + \Phi_{sl} = -\iint_{(S)} \vec{D}_d \cdot \vec{n}\, dS + \iint_{(S)} \vec{D}_v \cdot \vec{n}\, dS + \Phi_{sl}$$

The divergence of the electric displacement vector is zero in vacuum, so, using the divergence theorem, the conservation of flux is written as follows:

$$\iiint_{(\tau)} (\vec{\nabla} \cdot \vec{D}_v)\, d\tau = -\iint_{(S)} \vec{D}_d \cdot \vec{n}\, dS + \iint_{(S)} \vec{D}_v \cdot \vec{n}\, dS + \Phi_{sl} = 0$$

That is to say:

$$\vec{D}_v \cdot \vec{n}\, dS - \vec{D}_d \cdot \vec{n}\, dS + d\Phi_{sl} = 0 \qquad [1.208]$$

When points M_v and M_d tend towards M, the lateral surface tends towards zero ($h \to 0$) such that $d\Phi \to 0$. At the limit of point M, equation [1.208] gives the following equality:

$$\vec{D}_v \cdot \vec{n}\, dS = \vec{D}_d \cdot \vec{n}\, dS \Rightarrow D_{vn} = D_{dn} \qquad [1.209]$$

Result [1.209] expresses the continuity of the normal component of the electric displacement vector across a vacuum-dielectric surface of separation. What is it for the electric field? Using [1.207], we deduce from the first equality [1.209]:

$$(\varepsilon_0 \vec{E}_v)\cdot\vec{n}\,dS = (\varepsilon_0 \vec{E}_d + \vec{P})\cdot\vec{n}\,dS$$

That is to say:

$$\varepsilon_0 \vec{E}_v \cdot \vec{n}\,dS - \varepsilon_0 \vec{E}_d \cdot \vec{n}\,dS = \vec{P}\cdot\vec{n}\,dS \qquad [1.210]$$

According to equation [1.201], the surface density of polarization $\sigma_{pol} = \vec{P}\cdot\vec{n}$. Whereby, according to [1.199]:

$$\varepsilon_0 \vec{E}_v \cdot \vec{n}\,dS - \varepsilon_0 \vec{E}_d \cdot \vec{n}\,dS = \sigma_{pol}\,dS \Rightarrow E_{vn} - E_{dn} = \frac{\sigma_{pol}}{\varepsilon_0} \qquad [1.211]$$

The last equality [1.211] expresses that the normal component of the electric field has a discontinuity $\sigma_{pol}/\varepsilon_0$ across a vacuum-dielectric surface of separation. Now we need to study the behavior of the tangent components of the electric field and electric displacement vectors across the vacuum-dielectric surface of separation.

Let us consider a rectangular circuit (*C*) *ABDF* overlapping the surface of separation and whose lengths *AF* and *BD* are infinitesimal (Figure 1.30). We express the circulation of the electric field along the *ABDF* circuit using Stokes' theorem. As we only consider fields independent of time in dielectric media, the Maxwell-Faraday equation gives:

$$\vec{\nabla}\wedge\vec{E} = -\frac{\partial\vec{B}}{\partial t} = \vec{0}$$

With *dC* being the sum of circulations of the electric field along lengths *BD* and *FA*, we obtain:

$$\iint_{(S)} (\vec{\nabla}\wedge\vec{E}_v)\cdot\vec{dS} = \int_{(C)} (\vec{E}_v + \vec{E}_d)\cdot\vec{dl}' + dC = 0$$

That is to say:

$$\int_{(AB)} \vec{E}_v \cdot \vec{dl} - \int_{(DF)} \vec{E}_d \cdot \vec{dl} + dC = 0$$

Whereby:

$$(\vec{E}_v - \vec{E}_d)\cdot \vec{dl} + dC = 0 \Rightarrow \vec{E}_v \cdot \vec{dl} - \vec{E}_d \cdot \vec{dl} + dC = 0 \quad [1.212]$$

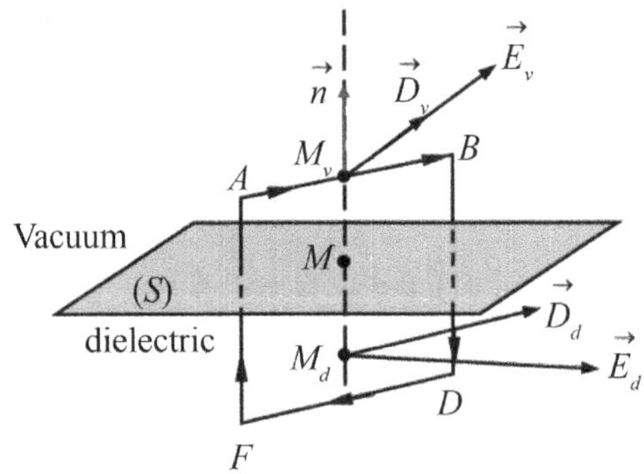

Figure 1.30. *Circulation of the electric field along an* ABDF *circuit overlapping a vacuum-dielectric surface of separation*

When M_v and M_d tend towards M, the lengths BD and FA tend towards zero such that $dC \to 0$. At point M, equation [1.211] gives the equality:

$$\vec{E}_v \cdot \vec{dl} - \vec{E}_d \cdot \vec{dl} = 0 \Rightarrow E_{vT} = E_{dT} \quad [1.213]$$

Result [1.213] expresses the continuity of the tangent component of the electric field across a vacuum-dielectric surface of separation. What is it for the displacement vector? Using [1.206], we deduce the following equation from [1.213]:

$$\frac{1}{\varepsilon_0}\vec{D}_v \cdot \vec{dl} - \frac{1}{\varepsilon_0}(\vec{D}_d \cdot \vec{dl} - \vec{P}\cdot \vec{dl}) = 0$$

That is to say:

$$\vec{D}_v \cdot \vec{dl} - \vec{D}_d \cdot \vec{dl} - \sigma_p \vec{n} \cdot \vec{dl} = 0 \quad [1.214]$$

At point M situated at the vacuum-dielectric surface of separation:

$$\vec{dl} = \overrightarrow{dAB} = -\overrightarrow{dAB} \Rightarrow \vec{dl}\cdot \vec{n} = 0$$

Equation [1.214] is written as follows:

$$\vec{D}_v \cdot \vec{dl} - \vec{D}_d \cdot \vec{dl} = 0 \Rightarrow D_{vT} = D_{dT} \qquad [1.215]$$

According to [1.215] the tangent component of the electric displacement vector is continuous across a vacuum-dielectric surface of separation.

1.2.2. *Homogeneous linear dielectric media*

1.2.2.1. *Definition of a dielectric linear medium*

The vacuum is characterized by its dielectric permittivity ε_0. A dielectric is characterized by its absolute dielectric permittivity ε and its relative permittivity ε_r linked to the permittivity of the vacuum via equation (which we will present below): $\varepsilon = \varepsilon_0\varepsilon_r$. For the vacuum: $\varepsilon_r = 1$.

By definition, a dielectric medium is said to be *homogeneous* if its properties are the same at all points. In this case, the equations between the physical parameters (fields, vectors, charge density, etc.) of the medium are independent of space coordinates. The dielectric is said to be *linear*, if the application passing from the electric displacement vector to the electric field is linear, i.e.:

$$\vec{D} = \varepsilon \vec{E} \qquad [1.216]$$

In general, for a non-homogeneous dielectric, the absolute dielectric permittivity ε depends on space coordinates x, y and z. It is a 3×3 matrix of matrix elements ε_{ij}. Equation [1.216] is also a matrix structure written in the following form:

$$(D) = (\varepsilon)(E) \Rightarrow \begin{pmatrix} D_x \\ D_y \\ D_z \end{pmatrix} = \begin{pmatrix} \varepsilon_{xx} & \varepsilon_{xy} & \varepsilon_{xz} \\ \varepsilon_{yx} & \varepsilon_{yy} & \varepsilon_{yz} \\ \varepsilon_{zx} & \varepsilon_{zy} & \varepsilon_{zz} \end{pmatrix} \begin{pmatrix} E_x \\ E_y \\ E_z \end{pmatrix} \qquad [1.217]$$

In the case of a homogeneous linear dielectric, the permittivity ε is independent of space coordinates x, y and z, which gives $D = \varepsilon E$.

1.2.2.2. *Matrix properties of dielectric permittivities, electric axes*

The dielectric permittivity matrix has special properties. Two matrix elements symmetrical with respect to the main diagonal are equal:

$$\varepsilon_{xy} = \varepsilon_{yx};\ \varepsilon_{xz} = \varepsilon_{zx};\ \varepsilon_{yz} = \varepsilon_{zy} \qquad [1.218]$$

Therefore the dielectric permittivity matrix is said to be *symmetrical.*

Moreover, the dielectric permittivity matrix is *diagonalizable*. There are three axes Mx_1, Mx_2 and Mx_3 perpendicular to one another at point M such that the matrix equation [1.217] is written:

$$\begin{pmatrix} D_1 \\ D_2 \\ D_3 \end{pmatrix} = \begin{pmatrix} \varepsilon_1 & 0 & 0 \\ 0 & \varepsilon_2 & 0 \\ 0 & 0 & \varepsilon_3 \end{pmatrix} \begin{pmatrix} E_1 \\ E_2 \\ E_3 \end{pmatrix} \qquad [1.219]$$

By definition, the axes Mx_1, Mx_2 and Mx_3 are called *electric axes* or *principal axes*. The parameters ε_1, ε_2 and ε_3 are called *principal dielectric permittivities* of M.

In addition, in the main axis system, the electric displacement vector is linked to the electric field by the second equation [1.207]. By omitting the index d set for dielectrics, we obtain:

$$\vec{D} = \varepsilon_0 \vec{E} + \vec{P} \qquad [1.220]$$

By projecting this equation on the main dielectric axes, considering the matrix relationship [1.219] we obtain:

$$\begin{cases} E_1 = \varepsilon_1 E_1 = \varepsilon_0 E_1 + P_1 \\ E_2 = \varepsilon_2 E_2 = \varepsilon_0 E_2 + P_2 \\ E_3 = \varepsilon_3 E_3 = \varepsilon_0 E_3 + P_3 \end{cases} \Rightarrow \begin{cases} P_1 = (\varepsilon_1 - \varepsilon_0) E_1 \\ P_2 = (\varepsilon_2 - \varepsilon_0) E_2 \\ P_3 = (\varepsilon_3 - \varepsilon_0) E_3 \end{cases} \qquad [1.221]$$

1.2.2.3. *Electric field and displacement vectors in a linear homogeneous isotropic dielectric (LHI), dielectric susceptibility*

We can distinguish between anisotropic dielectric media and isotropic dielectric media.

By definition, a dielectric medium is said to be *anisotropic* if its properties at a given point depend on the direction considered in space. This is the case, for example, of many crystals. In this case, the polarization vector $\vec{P}$ and the electric field $\vec{E}$ are not colinear as the previous equalities indicate [1.221]. However, the dielectric medium is linear and *isotropic* if its properties about a point are the same in all directions and at all points in the medium. In other words, for a linear and isotropic dielectric medium, the principal dielectric permittivities are equal. Dielectric liquids and non-crystalline solids, such as glass, are isotropic media. For electric field low values, the polarization vector is proportional to the electric field and given by the equation:

$$\vec{P} = \varepsilon_0 \chi_e \vec{E} \qquad [1.222]$$

By definition, χ_e is known as the *dielectric susceptibility*. Physically, χ_e is a measure of the separation of electric charges in a dielectric medium subjected to an external field $\vec{E}$. Using equation [1.222], the electric displacement vector [1.220] is written as follows:

$$\vec{D} = \varepsilon_0 (1+\chi_e)\vec{E} = \varepsilon \vec{E} \qquad [1.223]$$

In a linear and isotropic dielectric medium, the absolute dielectric permittivity is given by the equation:

$$\varepsilon = \varepsilon_0 (1+\chi_e) = \varepsilon_0 \varepsilon_r \ ; \varepsilon_r = (1+\chi_e) \qquad [1.224]$$

Generally speaking, χ_e is always greater than 1 ($\chi_e > 1$). According to the first equality [1.224], this means that $\varepsilon > \varepsilon_0$. Also, the last equality [1.224] shows that the relative permittivity ε_r is linked to the dielectric susceptibility χ_e. Therefore ε_r is also a measure of the degree of polarization of the dielectric medium. For vacuum, $\varepsilon_r = 1$ $\Rightarrow \chi_e = 0$.

Table 1.6 gives several values of relative permittivity for various dielectric media at ambient temperature.

Dielectric	ε_r	Dielectric	ε_r
Paper	2 – 2.8	Glass	7 – 8
Copper oleate Cu $(C_{18}H_{33}O_2)_2$	2.8	Alumina Al_2O_3	10.0
Acetamide C_2H_5NO	4.0	Copper sulfate $CuSO_4$	10.3
Barium nitrate $Ba(NO_3)_2$	5.9	Barium sulfate $BaSO_4$	11.4
Calcium carbonate $CaCO_3$	6.08	Copper oxide (I) Cu_2O	18.1
Ammonium chloride NH_4Cl	7.0	Methanol CH_4O	31.0
Calcium fluoride CaF_2	7.36	Crystallized sodium chloride NaCl	59.0

Table 1.6. *Relative permittivity of dielectric media at ambient temperature*

1.2.3. *Magnetic media*

1.2.3.1. *Magnetic medium, magnetization vector*

As we said in the introduction, charged particles in material media are in motion and are characterized by their magnetic spin and orbital moments. The action of an external magnetic field orientates elementary magnetic moments in well-defined directions. At the macroscopic scale, permanent magnetization of the material or induced by the action of an external magnetic field, causes the creation of a *magnetization vector* denoted $\vec{M}$. In a domain space with volume $d\tau$ of a magnetic material, the elementary magnetic moment $d\vec{M}$ is defined by equation:

$$\overrightarrow{dM} = \vec{M}\, d\tau \qquad [1.225]$$

The magnetization vector can be both uniform and not, in the elementary volume $d\tau$. It can also be static or vary with time.

1.2.3.2. *Vector potential in a magnetic medium, magnetization current densities*

Let us consider a magnetic medium occupying a certain space domain with volume τ (Figure 1.31). This domain is the site of a permanent or induced magnetization and is characterized by a distribution of currents due to conduction charge carriers. At point N in this domain, the magnetization vector is denoted $\vec{M}$ and the current density conduction vector is $\vec{J}_{free}$. The vector potentials $\overrightarrow{dA_c}$ and $\overrightarrow{dA_a}$ respectively created at point Q by the conduction current distribution $\vec{J}_{free}$ and by the magnetization of the medium characterized by the magnetization vector $\vec{J}$ are respectively deduced from the vector potential [1.70b] and the magnetic dipole potential [1.81] established above, i.e.:

$$\overrightarrow{dA_c} = \frac{\mu_0 \vec{J}_{free}\, d\tau}{4\pi r};\ \overrightarrow{dA_a} = \frac{\mu_0}{4\pi}\frac{\overrightarrow{dM} \wedge \vec{e}_r}{r^2} = \frac{\mu_0}{4\pi}\frac{\vec{M} \wedge \vec{e}_r}{r^2} d\tau \qquad [1.226]$$

The overall vector potential created in Q by the conduction current distribution and by the magnetization of the medium is given by the superposition principle, i.e.: $\overrightarrow{dA} = \overrightarrow{dA_c} + \overrightarrow{dA_a}$. For all conduction currents and magnetic moments distributed in volume τ, we obtain:

$$\vec{A} = \iiint_{(\tau)} \frac{\mu_0 \vec{J}_{free}\, d\tau}{4\pi r} + \iiint_{(\tau)} \frac{\mu_0}{4\pi}\frac{\vec{M} \wedge \vec{e}_r}{r^2} d\tau \qquad [1.227]$$

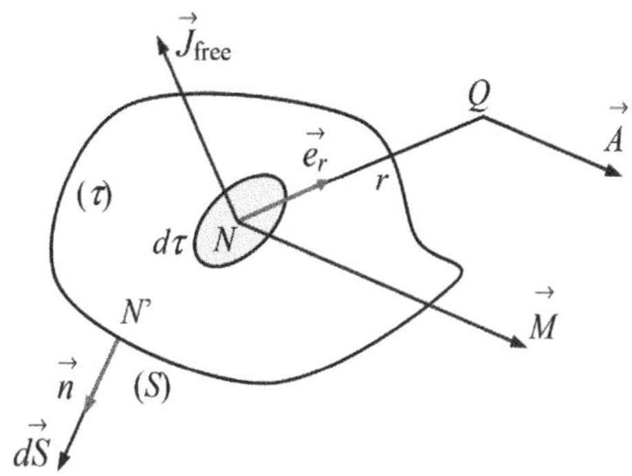

Figure 1.31. *Magnetic medium with volume τ*

Furthermore, due to the magnetization, it appears in volume currents distributed in volume (τ) and magnetization currents distributed on the surface (S) of the domain considered (Figure 1.31). These currents are known as *magnetization currents* and give rise to the vector potential $\overrightarrow{dA_a}$ created by the magnetization of the medium. We will attempt to expand on this.

As we can see, the overall vector potential [1.227] is the sum of a term in $1/r$ and another in $1/r^2$. To introduce the magnetization currents, let us transform the second term of the potential [1.227] to obtain an overall vector potential in $1/r$. Note equation [1.199a]:

$$\vec{\nabla}\left(\frac{1}{r}\right)=\frac{d}{dr}\left(\frac{1}{r}\right)\vec{e}_r=\frac{\vec{e}_r}{r^2}$$

Now, let us take advantage of the vector transformation equation:

$$\vec{\nabla}\wedge\left(\frac{\overrightarrow{M}}{r}\right)=\frac{1}{r}\vec{\nabla}\wedge\overrightarrow{M}+\vec{\nabla}\left(\frac{1}{r}\right)\wedge\overrightarrow{M} \qquad [1.228a]$$

Using [1.199a], the last equality [1.228a] gives:

$$\vec{\nabla}\wedge\left(\frac{\overrightarrow{M}}{r}\right)=\frac{\vec{\nabla}\wedge\overrightarrow{M}}{r}+\frac{\vec{e}_r\wedge\overrightarrow{M}}{r^2}=\frac{\vec{\nabla}\wedge\overrightarrow{M}}{r}-\frac{\overrightarrow{M}\wedge\vec{e}_r}{r^2}$$

So:

$$\frac{\vec{M} \wedge \vec{e}_r}{r^2} = \frac{\vec{\nabla} \wedge \vec{M}}{r} - \vec{\nabla} \wedge \left(\frac{\vec{M}}{r} \right) \qquad [1.228b]$$

If we add the last equality [1.228b] into [1.227], we obtain:

$$\vec{A} = \iiint_{(\tau)} \frac{\mu_0 \vec{J}_{free}\, d\tau}{4\pi r} + \iiint_{(\tau)} \frac{\mu_0}{4\pi} \frac{\vec{\nabla} \wedge \vec{M}}{r} d\tau - \iiint_{(\tau)} \frac{\mu_0}{4\pi} \vec{\nabla} \wedge \left(\frac{\vec{M}}{r} \right) d\tau \qquad [1.229a]$$

The first two terms of the right-hand side of [1.229a] are analogous to the condition $\vec{J}_{vmag} = \vec{\nabla} \wedge \vec{M}$. This allows the introduction of the magnetization volume current density vector $\vec{J}_{vmag}$. Equation [1.229a] is written as follows:

$$\vec{A} = \iiint_{(\tau)} \frac{\mu_0 \vec{J}_{free}\, d\tau}{4\pi r} + \iiint_{(\tau)} \frac{\mu_0}{4\pi} \frac{\vec{J}_{vmag}\, d\tau}{r} - \iiint_{(\tau)} \frac{\mu_0}{4\pi} \vec{\nabla} \wedge \left(\frac{\vec{M}}{r} \right) d\tau \qquad [1.229b]$$

Now we need to transform the third term of the right-hand part of expression [1.229b] to take account of the magnetization currents distributed on the surface. In order to apply the divergence theorem involving the divergence operator, it is first necessary to transform the third term to the right side of expression [1.229b]. To do this, let us take advantage of the following vector analysis property:

$$\vec{\nabla} \cdot (\vec{f} \wedge \vec{g}) = \vec{g} \cdot \vec{\nabla} \wedge \vec{f} - \vec{f} \cdot \vec{\nabla} \wedge \vec{g}$$

Substituting $\vec{f}$ for $\dfrac{\vec{M}}{r}$ we obtain:

$$\vec{\nabla} \cdot \left(\frac{\vec{M}}{r} \wedge \vec{g} \right) = \vec{g} \cdot \left(\vec{\nabla} \wedge \frac{\vec{M}}{r} \right) - \frac{\vec{M}}{r} \cdot \vec{\nabla} \wedge \vec{g} \qquad [1.229c]$$

This equality is true whatever the value of $\vec{g}$ and, in particular if $\vec{g}$ is constant. In this case, $\vec{\nabla} \wedge \vec{g} = \vec{0}$. Equation [1.229c] is then reduced to:

$$\vec{\nabla} \cdot \left(\frac{\vec{M}}{r} \wedge \vec{g} \right) = \vec{g} \cdot \left(\vec{\nabla} \wedge \frac{\vec{M}}{r} \right) \qquad [1.229d]$$

We introduce the term to the left of equation [1.229d] into the last term to the right of expression [1.229b]. Since $\vec{g}$ is constant, we get:

$$\vec{g}\cdot\iiint_{(\tau)}\frac{\mu_0}{4\pi}\vec{\nabla}\wedge\left(\frac{\overrightarrow{M}}{r}\right)d\tau=\iiint_{(\tau)}\frac{\mu_0}{4\pi}\vec{g}\cdot\vec{\nabla}\wedge\left(\frac{\overrightarrow{M}}{r}\right)d\tau$$

So, considering equality [1.229d], we obtain:

$$\vec{g}\cdot\iiint_{(\tau)}\frac{\mu_0}{4\pi}\vec{\nabla}\wedge\left(\frac{\overrightarrow{M}}{r}\right)d\tau=\iiint_{(\tau)}\frac{\mu_0}{4\pi}\vec{\nabla}\cdot\left(\frac{\overrightarrow{M}}{r}\wedge\vec{g}\right)d\tau \qquad [1.229e]$$

We can now apply the divergence theorem. For this we consider a point N' on the closed surface (S) surrounding the domain of volume (τ) (Figure 1.32, section 1.2.3.3). Here, $\vec{j}_{smag}=\vec{j}_{smag}(N')$ is the surface current density vector due to magnetization of the medium. According to the divergence theorem, the term to the right of equation [1.229e] is written as follows:

$$\frac{\mu_0}{4\pi}\iiint_{(\tau)}\vec{\nabla}\cdot\left(\frac{\overrightarrow{M}}{r}\wedge\vec{g}\right)d\tau=\frac{\mu_0}{4\pi}\iint_{(S)}\left(\frac{\overrightarrow{M}}{r}\wedge\vec{g}\right)\cdot\overrightarrow{dS}$$

That is to say:

$$\frac{\mu_0}{4\pi}\iiint_{(\tau)}\vec{\nabla}\cdot\left(\frac{\overrightarrow{M}}{r}\wedge\vec{g}\right)d\tau=\frac{\mu_0}{4\pi}\iint_{(S)}\vec{n}\cdot\left(\frac{\overrightarrow{M}}{r}\wedge\vec{g}\right)dS \qquad [1.229f]$$

So:

$$\vec{f}\cdot(\vec{h}\wedge\vec{k})=\vec{h}\cdot(\vec{k}\wedge\vec{f})=\vec{k}\cdot(\vec{f}\wedge\vec{h})$$

Considering this equation, we obtain:

$$\vec{n}\cdot\left(\frac{\overrightarrow{M}}{r}\wedge\vec{g}\right)=\vec{g}\cdot\left(\vec{n}\wedge\frac{\overrightarrow{M}}{r}\right). \qquad [1.229g]$$

Equation [1.229f] is therefore transformed as follows:

$$\frac{\mu_0}{4\pi}\iiint_{(\tau)}\vec{\nabla}\cdot\left(\frac{\overrightarrow{M}}{r}\wedge\vec{g}\right)d\tau=\frac{\mu_0}{4\pi}\iint_{(S)}\vec{g}\cdot\left(\vec{n}\wedge\frac{\overrightarrow{M}}{r}\right)dS \quad [1.229h]$$

Using equality [1.229d], equation [1.229h] gives:

$$\vec{g}\cdot\iiint_{(\tau)}\frac{\mu_0}{4\pi}\vec{\nabla}\wedge\left(\frac{\overrightarrow{M}}{r}\right)d\tau=\frac{\mu_0}{4\pi}\iint_{(S)}\vec{g}\cdot\left(\vec{n}\wedge\frac{\overrightarrow{M}}{r}\right)dS$$

Whereby:

$$\vec{g}\cdot\iiint_{(\tau)}\frac{\mu_0}{4\pi}\vec{\nabla}\wedge\left(\frac{\overrightarrow{M}}{r}\right)d\tau=\vec{g}\cdot\iint_{(S)}\frac{\mu_0}{4\pi}\left(\vec{n}\wedge\frac{\overrightarrow{M}}{r}\right)dS$$

So, after simplification:

$$\iiint_{(\tau)}\frac{\mu_0}{4\pi}\vec{\nabla}\wedge\left(\frac{\overrightarrow{M}}{r}\right)d\tau=\iint_{(S)}\frac{\mu_0}{4\pi}\left(\frac{\vec{n}\wedge\overrightarrow{M}}{r}\right)dS=-\iint_{(S)}\frac{\mu_0}{4\pi}\left(\frac{\overrightarrow{M}\wedge\vec{n}}{r}\right)dS$$

Taking this result into account, expression [1.229b] of the vector potential is written as follows:

$$\vec{A}=\iiint_{(\tau)}\frac{\mu_0\vec{J}_{free}\,d\tau}{4\pi r}+\iiint_{(\tau)}\frac{\mu_0}{4\pi}\frac{\vec{J}_{vmag}}{r}d\tau+\iint_{(S)}\frac{\mu_0}{4\pi}\left(\frac{\overrightarrow{M}\wedge\vec{n}}{r}\right)dS \quad [1.229i]$$

Now let us introduce the surface current density vector $\vec{j}_{smag}=\overrightarrow{M}\wedge\vec{n}$. The overall vector potential is written:

$$\vec{A}=\iiint_{(\tau)}\frac{\mu_0\vec{J}_{free}\,d\tau}{4\pi r}+\iiint_{(\tau)}\frac{\mu_0}{4\pi}\frac{\vec{J}_{vmag}\,d\tau}{r}+\iint_{(S)}\frac{\mu_0}{4\pi}\frac{\vec{J}_{smag}\,dS}{r} \quad [1.230]$$

Vector potential [1.230] is therefore created by:

– the conduction currents characterized by the current density vector: $\vec{J}_{free}$;

– the surface currents due to magnetization and characterized by the surface magnetization current density vector: $\vec{J}_{smag}=\overrightarrow{M}\wedge\vec{n}$;

– the volume currents due to magnetization and characterized by the volume magnetization current density vector: $\vec{J}_{vmag} = \vec{\nabla} \wedge \vec{M}$.

Thus, when the material medium considered has a permanent magnetization or is induced by an external magnetic field, the surface and volume magnetization current density vectors are given by the respective equations:

$$\vec{J}_{smag} = \vec{M} \wedge \vec{n}; \ \vec{J}_{vmag} = \vec{\nabla} \wedge \vec{M} \qquad [1.231]$$

When the magnetization disappears $(\vec{J}_{smag} = \vec{J}_{vmag} = \vec{0})$, only the current density conduction charges remain in the medium $\vec{J}_{free} = \vec{J}$. The corresponding potential is given by the first term of the general expression [1.230] as the Biot–Savart law expresses.

1.2.3.3. *Magnetic excitation field and vector*

To avoid any confusion, we will denote the current density conduction charges as $\vec{J}_{free} = \vec{J}$ to distinguish between the surface $\vec{J}_{smag}$ and volume $\vec{J}_{vmag}$ magnetization current density vectors.

Below is the Maxwell-Ampère equation deduced from Table 1.3:

$$\vec{\nabla} \wedge \frac{\vec{B}}{\mu_0} = \vec{J}_{free} + \varepsilon_0 \frac{\partial \vec{E}}{\partial t}$$

As we are interested in the field that is independent of time, this equation is written as follows:

$$\vec{\nabla} \wedge \frac{\vec{B}}{\mu_0} = \vec{J}_{free} \qquad [1.232]$$

Since the current density vector $\vec{J}_{free}$ is linked to the volume mobile charge density $(\vec{J}_{free} = \rho_{free}\vec{v})$, in a magnetized medium, we must consider the volume magnetization current density vector. Equation [1.232] is written as follows:

$$\vec{\nabla} \wedge \frac{\vec{B}}{\mu_0} = \vec{J}_{free} + \vec{J}_{vmag} \qquad [1.233]$$

By substituting the volume magnetization current density vector in this equation by expression [1.231], we obtain:

$$\vec{\nabla} \wedge \frac{\vec{B}}{\mu_0} = \vec{J}_{free} + \vec{\nabla} \wedge \vec{M}$$

That is to say:

$$\vec{\nabla} \wedge \frac{\vec{B}}{\mu_0} - \vec{\nabla} \wedge \vec{M} = \vec{J}_{free} \Rightarrow \vec{\nabla} \wedge \left(\frac{\vec{B}}{\mu_0} - \vec{M} \right) = \vec{J}_{free} \qquad [1.234]$$

That last of equalities [1.234] is analogous to equation [1.232]. This allows the introduction of a magnetic excitation vector $\vec{H}$ given by the equation:

$$\vec{H} = \frac{\vec{B}}{\mu_0} - \vec{M} \qquad [1.235]$$

In steady state, the magnetic excitation vector satisfies the local law according to [1.234]:

$$\vec{\nabla} \wedge \vec{H} = \vec{J}_{free} \qquad [1.236]$$

When the free charges are immobile, then $\vec{J}_{free} = \rho_{free} \vec{v} = \vec{0}$. Whereby, $\vec{\nabla} \wedge \vec{H} = \vec{0}$. The rotational of the excitation magnetic vector is non-zero if an electric current with intensity I circulates in the circuit considered. In other words, the rotational of the magnetic excitation vector is linked to the current intensity I.

To verify this, we will apply Stokes' theorem. Let us consider a closed contour (C) orientated as indicated in Figure 1.32. Here (S) is a surface based on the contour considered.

By applying Stokes' theorem, the circulation of the magnetic excitation vector on a closed contour (C) is:

$$\oint_{(C)} \vec{H} \cdot \vec{dl} = \iint_{(S)} (\vec{\nabla} \wedge \vec{H}) \cdot \vec{dS} \qquad [1.237]$$

Using [1.236], equation [1.237] becomes:

$$\oint_{(C)} \vec{H} \cdot \vec{dl} = \iint_{(S)} \vec{J}_{free} \cdot \vec{dS} = I \qquad [1.238]$$

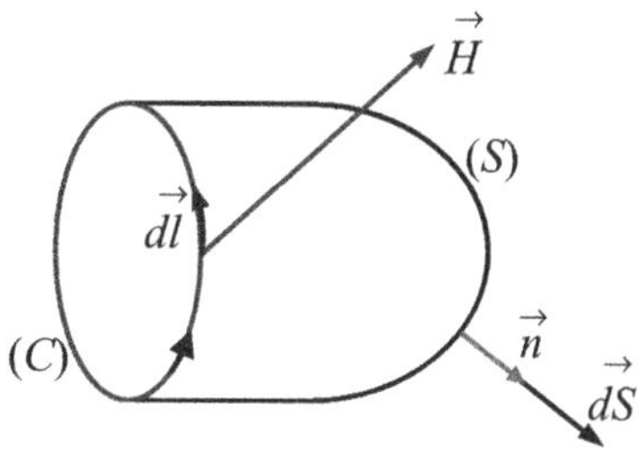

Figure 1.32. *Circulation of the excitation magnetic vector on a closed contour (C)*

Result [1.238] states that the circulation of the magnetic excitation vector on a closed contour (*C*) is equal to the intensity *I* of the current crossing the contour. Not that *I* is, by definition, the flux of the current density vector across a surface *dS* based on (*C*). So, equation [1.238] allows us to define the units of the magnetic excitation vector. Assuming *H* is constant, for any contour $Hl = I$, the magnetic excitation is expressed in ampere per meter ($A \cdot m^{-1}$) whereas the magnetic field is expressed in tesla (T).

1.2.3.4. *Continuity equations of the magnetic excitation field and vector crossing a vacuum-magnetic medium surface of separation*

Here we study the refraction of the magnetic field and the magnetic excitation vector crossing a vacuum-magnetic medium surface of separation. $\vec{B}_v$ and $\vec{H}_v$ and $\vec{B}_a$ and $\vec{H}_a$ denote respectively the magnetic field and the vector magnetic excitation in a vacuum and magnetized medium. According to [1.235]:

$$\vec{H}_v = \frac{\vec{B}_v}{\mu_0} \quad ; \quad \vec{H}_a = \frac{\vec{B}_a}{\mu_0} - \vec{M} \qquad [1.239]$$

In a vacuum, the magnetic excitation vector $\vec{H}_v$ is proportional to the magnetic field $\vec{B}_v$ [1.239]. Since $\mu_0 > 0$, then $\vec{H}_v$ and $\vec{H}_v$ have the same sense and direction. These are arbitrarily represented in Figure 1.33. However, in a magnetized medium, the magnetic excitation vector $\vec{H}_a$ and the magnetic field $\vec{B}_a$ do not have the same direction due to the magnetization vector $\vec{M}$. We choose to only represent the magnetic field $\vec{B}_a$.

Regardless of the current distribution considered, the divergence of the magnetic field is zero in accordance with the Maxwell-flux equation: $\vec{\nabla}\cdot\vec{B}=0$. If we consider a small cylinder with the base surface *dS*, center *N* and infinitesimal height *h* (Figure 1.33), we can express the total flux of the magnetic field leaving this cylinder. This flux is equal to the sum of the fluxes leaving the circular base surfaces *dS* with centers N_v and N_a and flux leaving the lateral surface of the cylinder.

Whereby:

$$\Phi_{tot}=\iint_{(S)}\vec{B}_a\cdot\vec{dS}+\iint_{(S)}\vec{B}_v\cdot\vec{dS}+\Phi_{sl}=-\iint_{(S)}\vec{B}_a\cdot\vec{n}\,dS+\iint_{(S)}\vec{B}_v\cdot\vec{n}\,dS+\Phi_{sl}$$

Using the divergence theorem, it becomes:

$$\iiint_{(\tau)}(\vec{\nabla}\cdot\vec{B}_v)d\tau=-\iint_{(S)}\vec{B}_a\cdot\vec{n}\,dS+\iint_{(S)}\vec{B}_v\cdot\vec{n}\,dS+\Phi_{sl}=0$$

That is:

$$\vec{B}_v\cdot\vec{n}\,dS-\vec{B}_a\cdot\vec{n}\,dS+d\Phi_{sl}=0 \qquad [1.240]$$

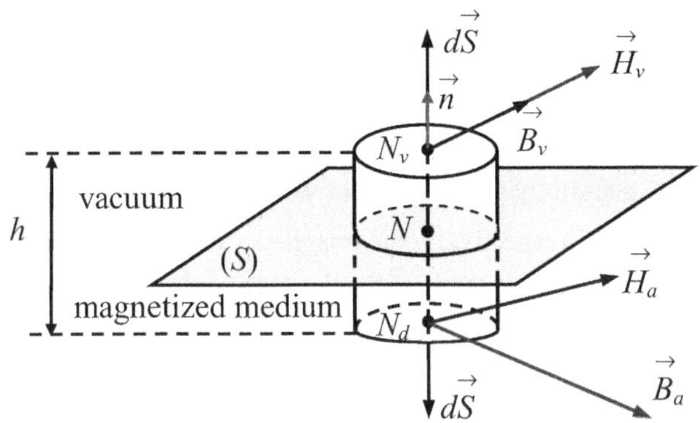

Figure 1.33. *Refraction of the magnetic field crossing a vacuum-magnetic medium surface of separation*

When N_v and N_d tend towards N, the lateral surface tends towards zero ($h \to 0$) such that $d\Phi \to 0$. At point N, equation [1.238] gives the equality:

$$\vec{B}_v \cdot \vec{n}\, dS = \vec{B}_a \cdot \vec{n}\, dS \Rightarrow B_{vn} = B_{an} \qquad [1.241a]$$

Result [1.241a] expresses the continuity of the normal component of the magnetic field crossing a vacuum-magnetic medium surface of separation. What about the magnetic excitation vector ?

Using [1.239], from the first equality [1.241] we deduce the equation:

$$(\mu_0 \vec{H}_v) \cdot \vec{n}\, dS = (\mu_0 \vec{H}_a + \vec{M}) \cdot \vec{n}\, dS \qquad [1.241b]$$

Using [1.231], we obtain:

$$\mu_0(\vec{H}_v \cdot \vec{n} - \vec{H}_a \cdot \vec{n}) = \vec{M} \cdot \vec{n} \Rightarrow \mu_0(H_{vn} - H_{an}) = M_n$$

So:

$$H_{vn} - H_{an} = \frac{M_n}{\mu_0} \qquad [1.241c]$$

According to [1.241c], the normal component of the magnetic excitation vector has a discontinuity M_n/μ_0 across a vacuum-magnetic medium surface of separation. Now let us study the behavior of tangent components of the magnetic field and magnetic excitation vector. Considering a rectangular circuit (*C*) *ABDF* overlapping the vacuum-magnetic medium surface of separation and infinitesimal lengths *AF* and *BD* (Figure 1.34).

We can simultaneously express the circulations of the magnetic field and the magnetic excitation vector along *ABDF* using Stokes' theorem. In a vacuum of currents, equation [1.233] is written as follows:

$$\vec{\nabla} \wedge \frac{\vec{B}}{\mu_0} = \vec{0}$$

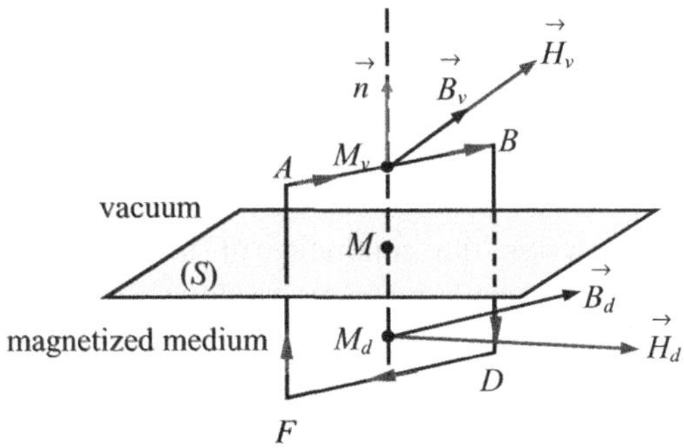

Figure 1.34. *Circulation of the magnetic field along an* ABDF *circuit overlapping a vacuum-magnetic medium surface of separation*

With dC being the sum of the considered circulations along lengths BD and FA, considering the result above, we obtain:

$$\iint_{(S)} \left(\vec{\nabla} \wedge \frac{\vec{B}}{\mu_0} \right) \cdot \overrightarrow{dS} = \int_{(C)} \left(\frac{\vec{B}_v}{\mu_0} + \frac{\vec{B}_a}{\mu_0} \right) \cdot \vec{dl}' + dC = 0$$

Therefore:

$$\int_{(AB)} \frac{\vec{B}_v}{\mu_0} \cdot \vec{dl} - \int_{(DF)} \frac{\vec{B}_a}{\mu_0} \cdot \vec{dl} + dC = 0$$

And ultimately:

$$\left(\frac{\vec{B}_v}{\mu_0} - \frac{\vec{B}_a}{\mu_0} \right) \cdot \vec{dl} + dC = 0 \Rightarrow \frac{\vec{B}_v}{\mu_0} \cdot \vec{dl} - \frac{\vec{B}_a}{\mu_0} \cdot \vec{dl} + dC = 0 \qquad [1.242]$$

When points M_v and M_d tend towards M, lengths BD and FA tend towards zero such that $dC \rightarrow 0$. At point M, equation [1.242] gives:

$$\frac{\vec{B}_v}{\mu_0} \cdot \vec{dl} - \frac{\vec{B}_a}{\mu_0} \cdot \vec{dl} = 0 \Rightarrow \frac{B_{vT}}{\mu_0} = \frac{B_{aT}}{\mu_0} = H_{vT} = H_{aT} \qquad [1.243]$$

The last equalities in [1.243] reflect the continuity of tangent components of the magnetic field and magnetic excitation vector crossing a vacuum-magnetic medium surface of separation.

1.2.3.5. *Linear and isotropic magnetic media, magnetic susceptibility*

Let us consider a magnetic medium with magnetic permeability $\mu = \mu\ (M)$. By definition, the magnetic medium is said to be *linear* and *isotropic*, if at any M in the medium, the application linking the magnetic excitation vector to the magnetic field is linear such that: μ = μ (M).

$$\vec{H} = \frac{\vec{B}}{\mu} \Rightarrow \vec{B} = \mu \vec{H} \qquad [1.244]$$

According to [1.235]:

$$\vec{B} = \mu_0 (\vec{H} + \vec{M}) \qquad [1.245]$$

By equalizing the last equality [1.244] and [1.245], we obtain:

$$\mu \vec{H} = \mu_0 \vec{H} + \mu_0 \vec{M} \Rightarrow \vec{M} = \frac{\mu - \mu_0}{\mu_0} \vec{H} \qquad [1.246]$$

Therefore:

$$\vec{J} = \chi_m \vec{H} \quad ; \quad \chi_m = \frac{\mu - \mu_0}{\mu_0} \qquad [1.247]$$

In equations [1.247], the physical variable χ_m is known as the *magnetic susceptibility* of the medium. Analogous to the relative dielectric permittivity ε_r, the *relative magnetic permeability* μ_r of the medium is defined by the equation:

$$\mu_r = \frac{\mu}{\mu_0} = 1 + \chi_m \qquad [1.248]$$

There are generally three types of magnetic media: paramagnetic, diamagnetic and ferromagnetic:

– *diamagnetism* is a property of all materials capable of interacting with a magnetic field and can be observed in copper (Cu), bismuth (Bi), gold (Au), silver (Ag), hydrogen

(H_2), etc. By subjecting a diamagnetic material to a non-uniform magnetic field, induced currents arise in the material and oppose the inductive field under Lenz's law. The resulting magnetic field is then reduced (the magnetic field lines become weak within the material and intense outside of it) and the material is moved outside the magnetic field. For a diamagnetic material:

$$\mu_r < 1 \quad ; \quad -10^{-4} < \chi_m < -10^{-9} \qquad [1.249]$$

The magnetic susceptibility χ_m is negative according to [1.249]: the magnetization vector $\vec{M} = \chi_m \vec{H}$ is opposite to the magnetic excitation vector $\vec{H}$;

– *paramagnetism* is a characteristic property of materials whose electronic layers are incomplete. When a paramagnetic material is subjected to a magnetic field, the uncompensated electronic magnetic moments align with the field. Paramagnetism can be observed in aluminum (Al), tungsten (W), platinum (Pt), tin (Sn), oxygen (O_2), etc. For a paramagnetic material:

$$\mu_r > 1 \quad ; \quad 10^{-6} < \chi_m < 10^{-4} \qquad [1.250]$$

The magnetic susceptibility χ_m is positive according to [1.250]: therefore, the magnetization vector $\vec{M} = \chi_m \vec{H}$ and the magnetic excitation vector $\vec{H}$ have the same sense. Unlike diamagnetic materials, magnetic field lines are more intense within a paramagnetic material than outside of it;

– *ferromagnetism* is a property of certain materials resulting in the alignment of the magnetic moments following the direction of the magnetic field applied in a domain known as a Weiss magnetic domain (region of a ferromagnetic material of uniform magnetization, with dimensions ranging between 10 μm and 1 m). For a ferromagnetic material:

$$\mu_r >> 1 \quad ; \quad \chi_m >> 1 \qquad [1.251]$$

Since the magnetic susceptibility χ_m is positive, the magnetization vector and the magnetic excitation vector have the same sense. In addition, the magnetic field lines are more intense inside the material than outside it. Note the existence of ferrimagnetism, which is a special case of ferromagnetism, as shown in Figure 1.35. For the vaccum, $\mu_r = 1 \Rightarrow \chi_m = 0$ according to [1.248].

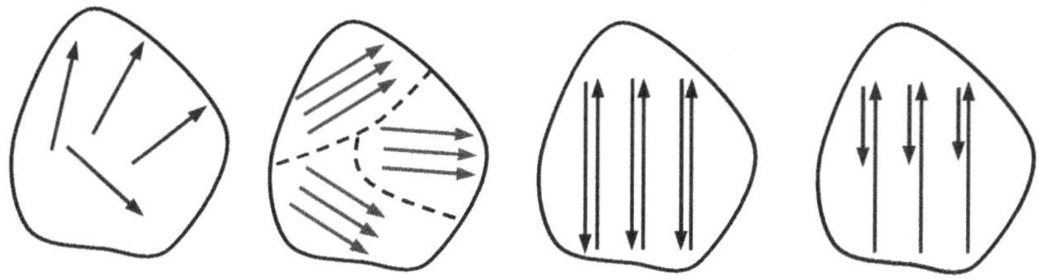

a) paramagnetism b) ferromagnetism c) anti - ferromagnetism d) ferrimagnetism

Figure 1.35. *Illustration of the different types of magnetism: a) magnetic moments distributed irregularly; b) magnetic moments aligned in a Weiss domain (10 μm to 1 m); c) antiparallel magnetic moments with equal intensities; d) antiparallel magnetic moments with different intensities*

Table 1.7 presents the relative permeability values for different magnetic materials at ambient temperature.

Magnetic material	μ_r (maximum value)
Cobalt (Co)	200
Nickel (Ni)	600
Iron (Fe)	5,000

Table 1.7. *Relative permeability of magnetic materials at ambient temperature*

1.2.4. *Maxwell equations in a polarized and magnetic medium*

1.2.4.1. *Couples of Maxwell's equations in a polarized dielectric medium*

Let us take the two pairs of Maxwell's equations grouped in Table 1.4 as well as the continuity equation [1.153] expressing the conservation of electric charge. We obtain the results as shown in Table 1.8 below.

First couple	$\vec{\nabla}\wedge\vec{E}=-\frac{\partial\vec{B}}{\partial t}$	Maxwell-Faraday
	$\vec{\nabla}\cdot\vec{B}=0$	Maxwell-flux
Second couple	$\vec{\nabla}\wedge\frac{\vec{B}}{\mu_0}=\vec{J}_{free}+\varepsilon_0\frac{\partial\vec{E}}{\partial t}$	Maxwell-Ampère
	$\vec{\nabla}\cdot\vec{E}=\frac{\rho_{free}}{\varepsilon_0}$	Maxwell-Gauss

Table 1.8. *Maxwell equations couples*

To establish the Maxwell equations in a polarized dielectric medium, we must take into account the electric displacement vector verifying [1.206]:

$$\vec{D}=\varepsilon_0\vec{E}+\vec{P}\Rightarrow\varepsilon_0\vec{E}=\vec{D}-\vec{P} \quad [1.252]$$

We claim that the first pair of Maxwell's equations is valid both in a vacuum and in a dielectric. We now need to modify the equation couple. Using [1.252], we obtain:

– *Maxwell-Ampère*:

$$\vec{\nabla}\wedge\frac{\vec{B}}{\mu_0}=\vec{J}_{free}+\varepsilon_0\frac{\partial\vec{E}}{\partial t}=\vec{J}_{free}+\frac{\partial\varepsilon_0\vec{E}}{\partial t}=\vec{J}_{free}+\frac{\partial(\vec{D}-\vec{P})}{\partial t} \quad [1.253a]$$

So:

$$\vec{\nabla}\wedge\frac{\vec{B}}{\mu_0}=\vec{J}_{free}+\frac{\partial\vec{D}}{\partial t}-\frac{\partial\vec{P}}{\partial t} \quad [1.253b]$$

– *Maxwell-Gauss*: to modify the Maxwell-Gauss equation, we use the local formula of Gauss's theorem [1.25] considering the dielectric medium, with a charge volume density ρ_{free} and a charge volume density of polarization ρ_{pol}

$$\vec{\nabla}\cdot\vec{E}=\frac{\rho_{free}+\rho_{pol}}{\varepsilon_0};\quad \rho_{pol}=-\vec{\nabla}\cdot\vec{P} \quad [1.254]$$

which gives:

$$\vec{\nabla}\cdot\vec{E}=\frac{\rho_{free}-\vec{\nabla}\cdot\vec{P}}{\varepsilon_0}\Rightarrow\varepsilon_0(\vec{\nabla}\cdot\vec{E})=\rho_{free}-\vec{\nabla}\cdot\vec{P}$$

therefore:

$$\varepsilon_0(\vec{\nabla}\cdot\vec{E})+\vec{\nabla}\cdot\vec{P}=\rho_{free}\Rightarrow\vec{\nabla}\cdot(\varepsilon_0\vec{E}+\vec{P})=\rho_{free}$$

And by considering [1.252]:

$$\vec{\nabla}\cdot\vec{D}=\rho_{free}$$

This corresponds to the local equation [1.205] verified by the electric displacement vector.

So, in a dielectric medium whose polarization is independent of time, the second Maxwell equation couple is written as follows:

$$\vec{\nabla}\wedge\frac{\vec{B}}{\mu_0}=\vec{J}_{free}+\frac{\partial\vec{D}}{\partial t};\ \vec{\nabla}\cdot\vec{D}=\rho_{free} \qquad [1.255]$$

In Table 1.9, we summarize Maxwell's equations in a dielectric with a steady state polarization.

First couple	$\vec{\nabla}\wedge\vec{E}=-\frac{\partial\vec{B}}{\partial t}$	$\vec{\nabla}\cdot\vec{B}=0$
Second couple	$\vec{\nabla}\wedge\frac{\vec{B}}{\mu_0}=\vec{J}_{free}+\frac{\partial\vec{D}}{\partial t}$	$\vec{\nabla}\cdot\vec{D}=\rho_{free}$

Table 1.9. *Maxwell's equations in a polarized dielectric medium*

Considering Table 1.9, we notice that like in a vacuum, Maxwell's equations are asymmetrical. The divergence of the magnetic field is zero while the divergence of the electric displacement vector is non-zero. The displacement vector flux is not conserved in a charged electric environment.

1.2.4.2. *Maxwell's equations in a polarized and magnetized material medium*

In a polarized dielectric medium, the *modified* Maxwell-Ampère equation is written as follows:

$$\vec{\nabla} \wedge \frac{\vec{B}}{\mu_0} = \vec{J}_{free} + \frac{\partial \vec{D}}{\partial t} \quad [1.256]$$

In a polarized and magnetic material medium, we must consider the magnetization volume density vector $\vec{J}_{vmag} = \vec{\nabla} \wedge \vec{M}$:

$$\vec{\nabla} \wedge \frac{\vec{B}}{\mu_0} = \vec{J}_{free} + \frac{\partial \vec{D}}{\partial t} + \vec{\nabla} \wedge \vec{M} \Rightarrow \vec{\nabla} \wedge \frac{\vec{B}}{\mu_0} - \vec{\nabla} \wedge \vec{M} = \vec{J}_{free} + \frac{\partial \vec{D}}{\partial t}$$

whereby:

$$\vec{\nabla} \wedge \left(\frac{\vec{B}}{\mu_0} - \vec{M} \right) = \vec{J}_{free} + \frac{\partial \vec{D}}{\partial t} \quad [1.257a]$$

By definition, in a polarized and magnetic material medium, the magnetic excitation vector is given by the expression:

$$\vec{H} = \frac{\vec{B}}{\mu_0} - \vec{M} \Rightarrow \vec{B} = \mu_0 (\vec{H} + \vec{M}) \quad [1.257b]$$

So, in a polarized and magnetic material medium, the *modified* Maxwell-Ampère equation is written as follows:

$$\vec{\nabla} \wedge \vec{H} = \vec{J}_{free} + \frac{\partial \vec{D}}{\partial t} \quad [1.258]$$

In Table 1.10, we summarize Maxwell's equations in a polarized and magnetic material medium.

First couple	$\vec{\nabla} \wedge \vec{E} = -\frac{\partial \vec{B}}{\partial t}$	$\vec{\nabla} \cdot \vec{B} = 0$
Second couple	$\vec{\nabla} \wedge \vec{H} = \vec{J}_{free} + \frac{\partial \vec{D}}{\partial t}$	$\vec{\nabla} \cdot \vec{D} = \rho_{free}$

Table 1.10. *Maxwell's equations in a polarized and magnetic material medium*

Note the presence of a magnetic current distribution $\vec{J}_{mag}$. In the context of diffraction theory, an equivalent magnetic charge ρ_{mag} is introduced to make Maxwell's equations more symmetrical (see the last column of Table 1.10). Thus, we modify the Maxwell-flux equation to transform a Maxwell-Thomson equation given by the equation:

$$\vec{\nabla} \cdot \vec{B} = \rho_{mag} \quad [1.259]$$

Moreover, the magnetic current vector $\vec{J}_{mag}$ and the magnetic charge ρ_{mag} would satisfy the continuity equation:

$$\vec{\nabla} \cdot \vec{J}_{mag} + \frac{\partial \rho_{mag}}{\partial t} = 0 \quad [1.260]$$

However, magnetic magnitudes $\vec{J}_m$ and ρ_m have no physical existence theoretically demonstrated to date. Maxwell's equations in a magnetic medium therefore remain equal to those grouped in Table 1.10.

1.2.4.3. *Integral forms of Maxwell's equations in a material medium*

Depending on the problem to be addressed, Maxwell's equations are used from their local expressions or in integral form as with the moment method or other methods based on finite elements. By considering a point N on surface (S) surrounding a polarized medium with volume τ. Where $dQ_{\text{int}} = \rho_{\text{free}}\, d\tau$ the free charge contained in the elementary volume $d\tau$. Using the divergence theorem:

$$\iint_{(S)} \vec{D} \cdot d\vec{S} = \iiint_{(\tau)} (\vec{\nabla} \cdot \vec{D}) d\tau$$

The Maxwell-Gauss equation is written as follows:

$$\iiint_{(\tau)} (\vec{\nabla} \cdot \vec{D}) d\tau = \iiint_{(\tau)} \rho_{free}\, d\tau = Q_{\text{int}} \qquad [1.261]$$

Equation [1.261] expresses the electric charge contained in volume τ and the vector field flux $\vec{D}$ across surface (S).

Now let us consider a polarized and magnetic medium. Using the Maxwell-Ampère equation, we express the circulation of the magnetic excitation vector along a closed contour (C) on which surface (S) is based.

We obtain:

$$\oint_{(C)} \vec{H} \cdot d\vec{l} = \iint_{(S)} \left(\vec{J}_{tot} + \frac{\partial \vec{D}}{\partial t} \right) \cdot d\vec{S} \qquad [1.262]$$

This integral form of the Maxwell-Ampère equation expresses the circulation of the vector field $\vec{H}$ along contour (C) and flux across surface (S) of the vector field $\left(\vec{J}_{total} + \frac{\partial \vec{D}}{\partial t} \right)$.

Now let us express the circulation of the electric field along a closed contour (C) on which surface (S) is based. We obtain from the integration of the Maxwell-Faraday equation:

$$\oint_{(C)} \vec{E} \cdot d\vec{l} = -\iint_{(S)} \frac{\partial \vec{B}}{\partial t} d\vec{S} \qquad [1.263]$$

Thus, since the divergence of the magnetic field is zero, then the integral form of the Maxwell-flux equation is written as follows:

$$\oiint_{(S)} \vec{B} . d\vec{S} = 0 \qquad [1.264]$$

This integral form expresses the fact that the flux of the magnetic field through a closed surface (Σ) is always zero. The integral forms of the Maxwell equations in material media are summarized in Table 1.11.

First couple	$\oint_{(C)} \vec{E} \cdot d\vec{l} = -\iint_{(S)} \frac{\partial \vec{B}}{\partial t} d\vec{S}$	Maxwell-Faraday
	$\oiint_{(S)} \vec{B}.d\vec{S} = 0$	Maxwell-flux
Second couple	$\oint_{(C)} \vec{H} \cdot d\vec{l} = \iint_{(S)} \left(\vec{J}_{tot} + \frac{\partial \vec{D}}{\partial t} \right) \cdot d\vec{S}$	Maxwell-Ampère
	$\iiint_{(\tau)} (\vec{\nabla} \cdot \vec{D}) d\tau = \iiint_{(\tau)} \rho_{free} \, d\tau = Q_{\text{int}}$	Maxwell-Gauss

Table 1.11. *Integral forms of Maxwell's equations in a material medium*

1.3. References

Amzallag, E., Cipriani, J., Aïm, J.B., Piccioli, N. (2006). *Électrostatique et électrocinétique*. Dunod, Paris.

Annequin, R., Boutigny, J. (1974). *Électricité 2*. Vuibert, Paris.

Bauduin, J.-M., Guérillot, A. (2001). *Électromagnétisme*. Ellipses, Paris.

Benson, H. (2015). *Électricité et magnétisme*. De Boeck, Brussels.

Bok, J., Hulin-Jung, N. (1979). *Relativité, ondes électromagnétiques*. Hermann, Paris.

Bruneaux, J., Saint-Jean, M., Matricon, J. (2002). *Électrostatique et magnétostatique*. Belin, Paris.

Faroux, J.-P., Renault, J. (1988). *Électromagnétisme*. Dunod, Paris.

Krempf, P. (2004). *Électromagnétisme MP*. Bréal, Paris.

Mauras, D. (1999). *Électromagnétisme*. PUF, Paris.

Sakho, I. (2018). *Électrostatique, magnétostatique et induction électromagnétique. Résumés de cours et 120 exercices corrigés*. Ellipses, Paris.

Stöcker, H., Junt, F., Guillaume, G. (2007). *Toute la physique*. Dunod, Paris.

2

The Propagation of Optical and Radio Electromagnetic Waves

Hervé SIZUN
ArmorScience, Lannion, France

2.1. Introduction

An electromagnetic wave is the propagation of an electrical disturbance of the matter: a static charge creates an electric field, a charge that moves creates a magnetic field, a disturbance in both creates an electromagnetic field that moves in space (vacuum, air, materials, etc.), that is an electromagnetic wave.

The parameters characterizing the propagation of electromagnetic waves are the electric field $\vec{E}$, the magnetic field $\vec{H}$, the electrical induction $\vec{D}$ and the magnetic induction $\vec{B}$ (Jouget 1978).

Only the vectors $\vec{E}$ and $\vec{B}$ produce actions through which it is possible to measure the electromagnetic field. The vectors $\vec{D}$ and $\vec{B}$ are linked to vectors $\vec{E}$ and $\vec{H}$ by the following linear equations:

$$\vec{D} = \varepsilon \vec{E} \qquad [2.1]$$

$$\vec{B} = \mu \vec{H} \qquad [2.2]$$

For a color version of all figures in this book, see www.iste.co.uk/favennec/electromagnetic1.zip.

Electromagnetic Waves 1,
coordinated by Pierre-Noël FAVENNEC.

The coefficients ε and μ depend on the medium in which the electromagnetic wave propagates. For homogeneous and isotropic media, ε and μ are constants. These are respectively known as the dielectric (or permittivity) constant and the magnetic permeability constant.

In a vacuum, they have the following values:

$$\varepsilon_0 = 10^{-9} / 36\pi = 8.842 \times 10^{-12} \qquad [2.3]$$

$$\mu_0 = 4\pi \times 10^{-7} \qquad [2.4]$$

The permittivity ε_0 and the magnetic permeability μ_0 values in a vacuum are respectively expressed in Farad/meter (F/m) and Henry/meter (H/m).

In a material medium, it is typical to link these coefficients to those of a vacuum by the following equations:

$$\varepsilon = \varepsilon_0 \times \varepsilon_r \qquad [2.5]$$

$$\mu = \mu_0 \times \mu_r \qquad [2.6]$$

where ε_r and μ_r are respectively the permittivity and the relative magnetic permeability of the medium.

The permittivity and the magnetic permeability characterize the refractive index of the medium defined by the following equation:

$$n = \sqrt{\varepsilon\mu} \qquad [2.7]$$

When there are charged particles in motion (with density ρ) in space, this transfer of charges creates an electric current which is characterized by its density, vector $\vec{J}$. The transfer of electric charges, described by vector $\vec{J}$, can simply be the result of the effect of an electric field $\vec{E}$. In this case, $\vec{J}$ is proportional to $\vec{E}$ and for a homogeneous and isotropic medium, we obtain:

$$\vec{J} = \sigma\vec{E} \qquad [2.8]$$

The coefficient σ is the electrical conductivity of the medium. It is measured in Mhos per meter or Siemens per meter (Sm^{-1}).

NOTE.– The transfer of charges can also occur due to an external cause entirely independent to the electromagnetic field (for example, the action of a generator with its own energy source). In this case, it affects vector $\vec{J}$ with index 0. Also, in the general case we have the following:

$$\vec{J} = \vec{J}_0 + \sigma\vec{E} \qquad [2.9]$$

2.2. Maxwell's equations

The vast world of electromagnetism is described by Maxwell's four equations which mathematically model the interactions between electric charges, electric currents, electric fields and magnetic fields. They associate the following five parameters that define the electromagnetic state of a medium: $\vec{E}$, $\vec{B}$, $\vec{D}$, $\vec{H}$ and $\vec{J}$.

2.2.1. *Maxwell-Gauss equation*

A charged body or particle is a concentration of electric charges with the same sign. This causes an electric field $\vec{E}$ to appear in the surrounding space.

The Maxwell-Gauss equation states that the divergence of the electric field is proportional to the distribution of electric charges:

$$\vec{\nabla}.\vec{E} = div\vec{E} = \frac{\rho}{\varepsilon_0} \qquad [2.10]$$

where:

– ρ is the density of electric charges;

– ε_0 is the permittivity in a vacuum.

For more details concerning the mathematical tools used in the remainder of this chapter (gradient, divergence, rotational, Laplacian, etc.), the reader should refer to Appendix 2 on vector calculations.

The density of electric charges is measured in Coulomb/m (linear density), Coulomb/m^2 (surface density) or Coulomb/m^3 (volumetric density), according to whether the charges are distributed linearly, across a surface or in a volume in the MKSA international system (meter, kilogram, second, ampere).

The permittivity of a vacuum or dielectric constant is expressed in H/m (Henry per meter).

According to the Maxwell-Gauss equation, the electric field is divergent or convergent according to the sign of the charge from the source and is proportional to the distribution of these charges. The electric field lines begin in the center and diverge towards infinity (Figure 2.1). The greater the charges, the more intense the field.

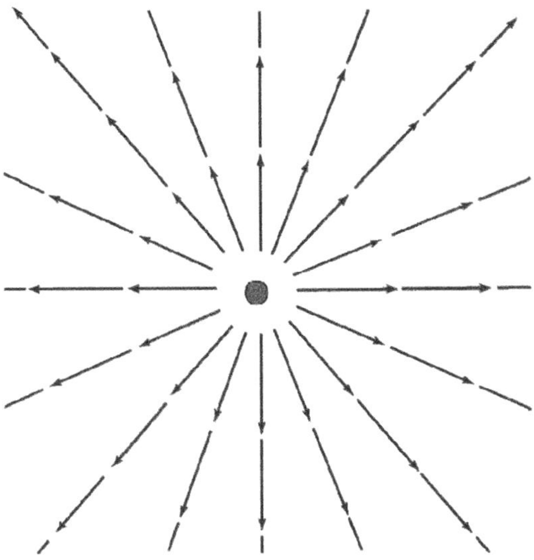

Figure 2.1. *Configuration of field lines of the electric field*

2.2.2. *Maxwell-Thompson equation*

The divergence of the magnetic field is zero:

$$\vec{\nabla}.\vec{B} = div\vec{B} = 0 \qquad [2.11]$$

The magnetic field is not divergent: the magnetic field lines do not go to infinity. The field lines go from one pole to another (Figure 2.2).

There is no magnetic monopole unlike the existence of an electric monopole such as an electron, which is negative, or a proton, which is positive. If we split a magnet into two parts, we obtain two magnets each with a north pole and a south pole.

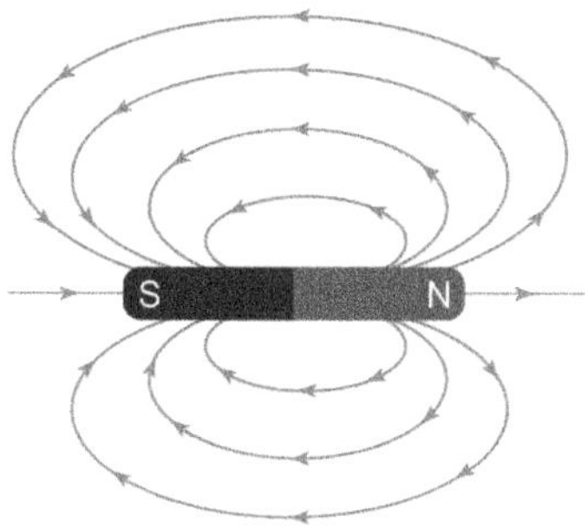

Figure 2.2. *Configuration of magnetic field lines*

2.2.3. *Maxwell-Faraday equation*

The rotational of the electric field is proportional to the variation in the magnetic field with time:

$$\vec{\nabla} \wedge \vec{E} = -\frac{\partial \vec{B}}{\partial t} \tag{2.12}$$

It is the variation in magnetic field (and not the magnetic field itself) that produces an electric field.

If we place a magnet in a coil, nothing happens. However, if you twist and turn the magnet, it creates an electric field around it, which in turn generates an electric current in the wire. For example, the dynamo of a bicycle only provides power to its lights when it moves.

This dependence of electrical phenomena on the variation in magnetic field was discovered by Faraday.

2.2.4. *Maxwell-Ampère equation*

The rotational of the magnetic field is the sum of its dependence on the variation in electric field over time and on a fixed electric current:

$$\vec{\nabla} \wedge \vec{B} = \mu_0 \vec{J} + \varepsilon_0 \mu_0 \frac{\partial \vec{E}}{\partial t} \tag{2.13}$$

The magnetic field is due to the variation in electric field over time. It also depends on an electric current in the case of an electrical conductor.

The current does not need to vary for there to be a magnetic field, whereas the magnetic field must vary to generate an electric current.

The Maxwell-Faraday and Maxwell-Ampère equations show that both electric and magnetic fields are coupled and that the variation in one is proportional to the intensity of the other's field.

They describe the conversion of the magnetic component of an electromagnetic wave into its electric component and vice versa.

An electromagnetic wave can therefore propagate without any other support than from itself.

If the term $\mu_0 \vec{J}$ is removed from the Maxwell-Ampère equation, we find it is symmetrical to the Maxwell-Faraday equation.

2.3. Solving Maxwell's equations

Solving Maxwell's equations in free space leads us to the Helmholtz equation for each of the following vectors: $\vec{E}$, $\vec{B}$, $\vec{D}$ and $\vec{H}$.

For field $\vec{E}$ we have the following equation:

$$\left(\Delta - \varepsilon\mu \frac{\partial^2}{\partial t^2}\right)\vec{E} = 0 \quad [2.14]$$

where Δ is the Laplacian operator: grad(div) – rot (rot) (see Appendix 2).

By limiting ourselves to a purely sinusoidal solution, it can be written in the following form:

$$E = E_0 \exp\left[j\left(\omega t - kz\right)\right] \quad [2.15]$$

where:

- E_0 is the amplitude of the field;
- $\omega = 2\pi f$ is the angular frequency;

– t is the elapsed time;

– k is the wave number;

– z is the distance traveled along axis z.

The wave number represents the rate of variation in the field phase at distance z. This is a magnitude proportional to the number of oscillations that a wave travels per unit length. The wave phase has a kz radian variation over a distance of z meters.

At the surface of separation of two media with different dielectric characteristics, we have the continuity of tangent component of the electric and magnetic fields and the continuity of normal components of electrical and magnetic inductions.

2.4. Characteristics of electromagnetic waves

2.4.1. *Propagation speed*

The Helmholtz equation shows that the oscillations of vector $\vec{E}$ in constant phase propagate in space in the form of a wave with a propagation factor γ such that:

$$\gamma^2 = \varepsilon\mu \frac{\partial^2}{\partial t^2} \qquad [2.16]$$

The expression of γ^2 for a non-absorbent medium is equal to:

$$\gamma^2 = \frac{1}{v^2}\frac{\partial^2}{\partial t^2} \qquad [2.17]$$

By comparing the equations, we see that the propagation rate of electromagnetic waves is therefore equal to:

$$v = \frac{1}{\sqrt{\varepsilon\mu}} \qquad [2.18]$$

In a vacuum we have:

$$\frac{1}{\sqrt{\varepsilon_0\mu_0}} = c = 3\times10^8 m/s \qquad [2.19]$$

where constant c is the speed of light.

2.4.2. *Wavelength and/or frequency*

Solving Helmholtz equations in the case of an electromagnetic plane wave shows that the electric field $\vec{E}$ and magnetic field $\vec{H}$ vectors are perpendicular and oscillate in phase in time and space.

Here we refer to a transverse electromagnetic (TEM) wave. The electric and magnetic fields propagate in phase, perpendicular to one another (Figure 2.3).

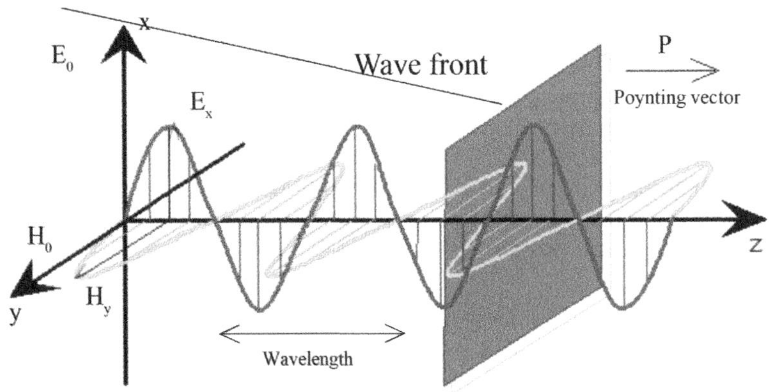

Figure 2.3. *Diagram depicting the propagation of an electromagnetic wave*

The wavelength and the frequency are linked by the equation:

$$\lambda = \frac{v}{f} = \frac{2\pi}{\omega\sqrt{\varepsilon\mu}} = vT \qquad [2.20]$$

where:

– f is the wave frequency in Hertz;

– ω is the pulsation;

– T is the period.

The wavelength represents the distance the wave phase varies by 2π.

The wavelength and/or the frequency characterize the electromagnetic wave, its propagation and its application domain.

Electromagnetic waves have different frequency ranges: ELF waves (f < 3 kHz), VLF waves (3 kHz < f < 30 kHz), LF waves (30 kHz < f < 300 kHz), MF waves (300 kHz < f < 3 MHz), HF waves (3 MHz < f < 30 MHz), VHF waves (30 MHz < f < 300 MHz), UHF waves (300 MHz < f < 3 GHz), SHF waves (3 GHz < f < 30 GHz), EHF waves (30 GHz < f < 300 GHz), infrared waves (3 THz < f < 430 THz) and optical waves (430 THz < f < 860 THz).

For more details on the terrestrial influences, the system considerations and the services associated to each, the reader should refer to Appendix 3 on frequency spectrum.

2.4.3. *The characteristic impedance of the propagation medium*

The ratio of amplitude of the electric field to the amplitude of the magnetic field is:

$$\frac{E_x}{H_y} = \left(\frac{\mu}{\varepsilon}\right)^{\frac{1}{2}} = Z_c \qquad [2.21]$$

We verify that this ratio has the physical dimensions of an electrical resistance. We call this the characteristic impedance of the propagation medium.

For a vacuum its value is:

$$Z_0 = \left(\frac{\mu_0}{\varepsilon_0}\right)^{\frac{1}{2}} = 377 Ohms \qquad [2.22]$$

NOTE.– The characteristic impedance does not only depend on the propagation medium, it also depends on the type of wave (TEM wave – transverse electromagnetic wave, spherical wave, guided wave). We therefore call Z_0 the wave impedance.

2.4.4. *Poynting vector*

Measured in Watts per square meter (W/m^2), the Poynting vector describes the amplitude and the direction of the flow of power transported by the wave per square meter of surface parallel to the plane (x, y), that is the power density. Its instant value is given by the following equation:

$$\vec{P} = \vec{E} \wedge \vec{H} \qquad [2.23]$$

Its average value over a given period, representing the actual power transported by the wave, is generally used. It is given by the following equation:

$$P_{average} = \frac{E_0 \times H_0}{2} \quad [2.24]$$

2.4.5. *The refractive index*

The refractive index plays an important role in the troposphere. The refractive index gradients in the vertical profile create guide layers for electromagnetic radiation. If these guide layers are wide enough, they can cause significant variations in the level of the direct signal, variations in angles of arrival and the appearance of multiple paths which interfere with the receiver.

It is defined by the following equation:

$$n = \sqrt{\varepsilon_r \mu_r} \quad [2.25]$$

The resulting propagation velocity is equal to:

$$v = \frac{1}{\sqrt{\varepsilon_0 \varepsilon_r \mu_0 \mu_r}} = \frac{1}{\sqrt{\varepsilon_0 \mu_0}} \times \frac{1}{\sqrt{\varepsilon_r \mu_r}} = \frac{c}{n} \quad [2.26]$$

In an absorbent medium, the refractive index can be written in the following complex form:

$$n = n' - in'' \quad [2.27]$$

The wave field becomes:

$$E = E_0 \exp\left[j\left(\omega t - nk_0 z\right)\right] \quad [2.28]$$

$$E = E_0 \exp\left[j\left(\omega t - \left(n' - in''\right)k_0 z\right)\right] \quad [2.29]$$

$$E = E_0 \exp\left(-n'' z\right)\exp\left[j\left(\omega t - n' k_0 z\right)\right] \quad [2.30]$$

The term exp(-n”z) represents the attenuation of the wave across the propagation medium.

The imaginary part of the refractive index represents the attenuation coefficient of the wave (α in m^{-1}). It depends on the permittivity, the permeability and the conductivity (Sm^{-1} or $(\Omega m)^{-1}$) of the medium, as well as the wave frequency.

The dielectric parameters such as permittivity, permeability and conductivity characterize the propagation media, the materials.

The field decreases exponentially as a function of distance. The distance over which the field decreases by a ratio of e (is divided by e) is known as the penetration depth (δ in m). It is given by the relation:

$$\delta = \frac{1}{\alpha} \quad [2.31]$$

2.4.6. *Polarization*

An electromagnetic wave has two active orthogonal components (the electric field vector and the magnetic field vector), which are in phase in time and space (Figure 2.4).

The polarization plane is defined by the direction of propagation (Oz) and that of the electric field (Ox). Figure 2.4 shows a vertically polarized wave. The polarization direction is determined by the direction in which it was generated. The use of vertical dipole antennas makes it possible to emit such a polarization.

However, if the electric field is horizontal, we refer to horizontally polarized transverse electromagnetic waves (TEM waves).

Since Maxwell's equations are linear, we can linearly combine several transverse electromagnetic waves:

$$E_x = E_1 \exp\left[j\left(\omega t - \beta z\right)\right] \quad [2.32]$$

$$E_y = E_2 \exp\left[j\left(\omega t - \beta z + \psi\right)\right] \quad [2.33]$$

where:

$$\beta = \frac{\omega}{v} = \frac{2\pi f}{c\lambda} \quad [2.34]$$

and ψ represents the phase difference of the second wave with respect to the first.

In a given wave plane (for example z = 0), the extremity of the resulting electric field $\vec{E} = \vec{E}_x + \vec{E}_y$ describes an ellipse whose equation is obtained in the parametric form by taking the real parts of both equations below and by making z = 0:

$$E_x = E_1 \cos(\omega t) \qquad [2.35]$$

$$E_y = E_2 \cos(\omega t + \psi) \qquad [2.36]$$

We call this wave an elliptically polarized plane wave.

When $E_1 = E_2$ and if $\Psi = \pi/2$, the ellipse transforms into a circle and the wave becomes circularly polarized.

Depending on the direction along which $\vec{E}$ rotates, the polarization can be left circular (levogyrous) or right circular (dextrogyrous) polarization.

Figure 2.4 illustrates these different types of polarization.

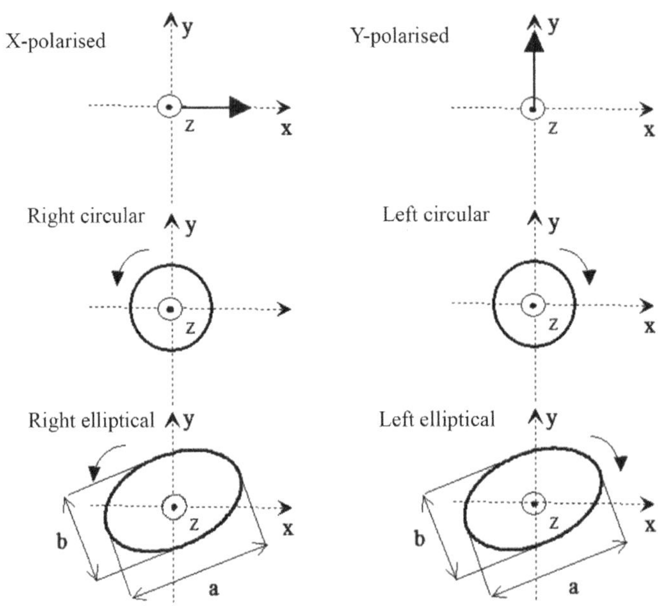

Figure 2.4. *The different polarization states for a wave propagating in direction z*

2.4.7. *Transpolarization*

In order to increase the transmission capacity on a given link, without increasing the bandwidth of the signal used, orthogonal rectilinear or circular polarizations are generally used. Unfortunately when passing through the atmosphere, mainly due to the presence of asymmetric raindrops and ice crystals, or even clear air, part of the energy radiated with a certain polarization is found on the orthogonal polarization thus causing interference between the two channels.

Depolarization is thus the phenomenon by which all or part of a radioelectric wave emitted with a given polarization no longer has any determined polarization after propagation. Transpolarization is the appearance, during propagation, of an orthogonal polarization component (cross-polar component) to the initial polarization (co-polar component). This phenomenon can be caused by rain or hydrometeors in the troposphere, multipath propagation, tropospheric scintillation or ionospheric scintillation.

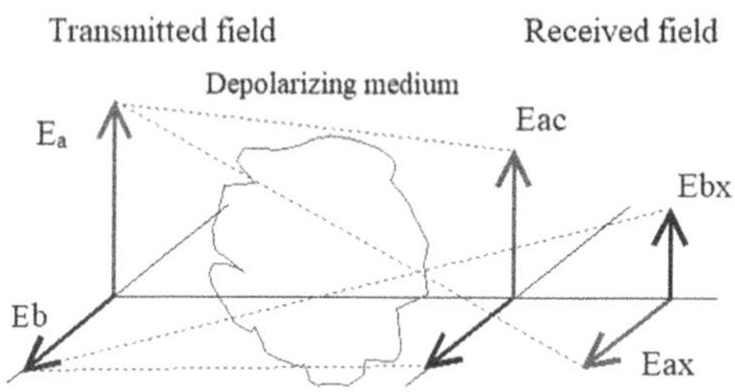

Figure 2.5. *Schematic of transpolarization*

Two parameters characterize this transpolarization: the polarization discrimination or the decoupling ratio XPD and the polarization isolation XPI.

2.4.7.1. *Polarization discrimination (XPD)*

The polarization discrimination or decoupling ratio (XPD) is the ratio, at the receiving point, of the power received with the initial polarization, to the power received with orthogonal polarization, and this for an electromagnetic wave emitted with a given polarization. It expresses how much a wave with a given polarization is found with orthogonal polarization. It depends on the propagation medium crossed by the wave. It is expressed in decibels:

$$XPD = 20\log\left(\frac{E_{ac}}{E_{ax}}\right) \qquad [2.37]$$

It is closely linked to the co-attenuation of the wave. Many models of this exist in the literature.

2.4.7.2. *Polarization isolation (XPI)*

When two orthogonally polarized components E_a and E_b are emitted at the same level, the ratio of copolar power E_{ac} (or E_{bc}) to contrapolar power E_{bx} (or E_{ax}) defines the polarization isolation XPI. These two ratios play an important role in the design of systems. The value of the ratio E_{ac}/E_{bx} is not necessarily the same as that of the ratio E_{bc}/E_{ax}. It expresses how much two orthogonally polarized signals, transmitted simultaneously, interfere at the reception. It depends on the propagation medium crossed by the wave. It is expressed in decibels:

$$XPI = 20\log\left(\frac{E_{ac}}{E_{bx}}\right) \qquad [2.38]$$

2.4.8. *Different propagation paths*

Mobile radio propagation, in particular, is a multi-path propagation (we also find this in ionospheric propagation with reflection on the different layers). Signals often follow paths of different lengths, travel times and arrive at the receiver in different phases.

Among these paths we distinguish:

– *direct paths*: the waves follow the line segment passing through the point of emission and reception;

– *transmitted paths*: generally, the direct route does not exist. It is often obstructed by buildings, vegetation, etc; it does not usually play a very important role;

– *reflected paths*: the waves arrive at the receiver following reflections on different obstacles (building facades, mountainsides, on the ground, etc.). These different paths contain enough energy to be efficient;

– *diffracted paths*: waves can bypass obstacles by diffraction (vegetation, buildings, roof edges, hills, etc.;

– *diffused paths*: this phenomenon is encountered more particularly when the wave encounters surfaces that are not smooth (soil, vegetation, sea, etc.) or small heterogeneities compared to the wavelength. The notion of a smooth surface is closely related to the frequency.

These different paths, with different amplitudes and phases, interfere when arriving at the receiver. Interference is either constructive (the paths arrive in phase) where there is signal reinforcement, or destructive, where there is attenuation of the signal. In addition, it should be noted that the mobile itself moves inside this figure of interferences. It successively detects bright and dark bands (interference fringes) causing attenuation of the signal.

2.4.9. *Fresnel zones*

The study of electromagnetic waves between an emitter E and a receiver R leads to the subdivision of the propagation space with a family of ellipsoids, called Fresnel ellipsoids, with points E and R being their focal points, such that any point M of one of these ellipsoids satisfies the relation (Figure 2.6):

$$EM + MR = ER + n\frac{\lambda}{2} \tag{2.39}$$

where:

– n is an integer that characterizes the ellipsoid considered (n = 1 characterizes the first Fresnel ellipsoid, etc.);

– λ is the wavelength.

As a general rule, the propagation is considered to be in the line of sight, that is to say with negligible diffraction phenomena, if there are no obstacles inside the 1st Fresnel ellipsoid. The latter delimits the region of space where almost all of the energy passes.

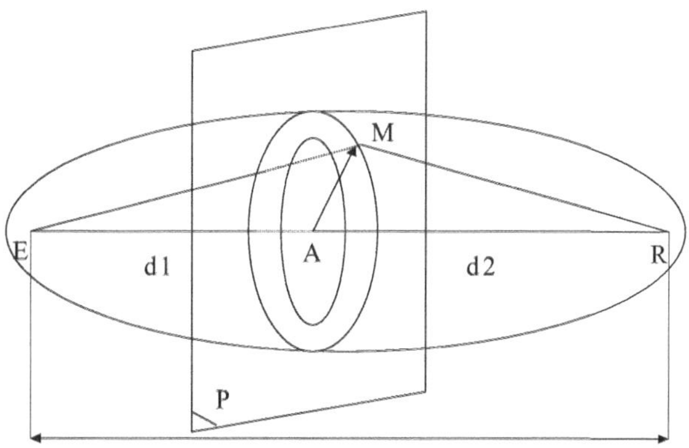

Figure 2.6. *Schematic representation of Fresnel zones*

The radius of the nth ellipsoid at a point on the path situated at a distance d_1 from E and a distance of d_2 from R ($d_1 + d_2 = d$, provided that d_1 and d_2 are large in relation to the radii of the ellipsoids) is given by the following equation:

$$R_n = \sqrt{\frac{n\lambda d_1 d_2}{d_1 + d_2}} \qquad [2.40]$$

The radius of the first ellipsoid is equal to:

$$r = \sqrt{\frac{\lambda d_1 d_2}{d_1 + d_2}} \qquad [2.41]$$

The maximum value at the middle of the path of this radius is:

$$r_{max} = \frac{1}{2}\sqrt{\lambda d} \qquad [2.42]$$

For example:

d = 1 km

f = 900 MHz, λ = 33 cm, r_{max} = 9 m

f = 1 800 MHz, λ = 16.5 cm, r_{max} = 6.4 m

2.4.10. *Fundamental properties of the propagation channel*

Due to the presence of multiple paths and the movement of the receiving station, the propagation channel has three fundamental properties: attenuation, variability and frequency selectivity.

The notion of attenuation was sufficient to study the propagation channel. In the context of digital communication, the fading of signals due to variability and selectivity induces degradations in the quality of the communication, independently of the attenuation.

2.4.10.1. *Propagation attenuation*

Different models of attenuation are defined in the literature. We generally distinguish the free-space attenuation and the attenuation relative to free-space (excess attenuation).

Free-space attenuation is the transmission attenuation due to the dispersion of the wave energy which would be obtained if the antennas were replaced by isotropic antennas, placed in a perfect, homogeneous and unlimited dielectric medium, with the distance between antennas being preserved. It is given by the following equation:

$$A_0 = \left(\frac{4\pi d}{\lambda}\right)^2 \quad [2.43]$$

In the logarithmic form it becomes:

$$A_0(dB) = 32.4 + 20\log_{10}(f) + 20\log_{10}(d) \quad [2.44]$$

where:

– A_0(dB) is the attenuation in decibels (see Appendix 4);

– d is the distance between the transmitter and the receiver (km);

– λ is the wavelength (km);

– f is the frequency (MHz).

For example:

d = 1 km

f = 900 MHz, A_0 = 91.4 dB

f = 1,800 MHz, A_0 = 97.4 dB

This assumes that only one radioelectric path has its first Fresnel ellipsoid cleared.

This is an expression in $1/d^2$ (energy conservation). This means that 6 dB is lost each time the distance is doubled.

The frequency dependence ($1/f^2$) depends on the configuration of the transmission (point-to-point propagation, radio-mobile propagation).

The excess attenuation, with regard to free-space attenuation, is the difference between propagation attenuation and free-space attenuation (absorption by gases, hydrometeors, walls, vegetation, diffraction attenuation, etc.).

2.4.10.1.1. Point-to-point links

The two antennas (transmitter and receiver) are orientated towards one another. The reasoning proceeds from the consideration of the volume of the antenna that is its equivalent area. The power received, as a function of the transmitted power P_e, is equal to:

$$P_r = P_t.A_e.A_r.f^2 \quad [2.45]$$

where:

- A_e is the equivalent area of the transmitting antenna;
- A_r is the equivalent area of the receiving antenna.

In this case there is an increase of 6 dB each time the frequency is doubled.

2.4.10.1.2. Radiomobile link

In radiomobile communication, the base station ignores the position of the mobile. The size of the receiving antennas becomes smaller and smaller as the frequency increases. Omnidirectional antennas are used and therefore not pointed ones. The power received is the product of the gains of the antennas (transmitter (G_e) and receiver (G_r)) by the power transmitted over f^2.

$$P_r = \frac{P_t.G_e.G_r}{f^2} \quad [2.46]$$

6 dB is lost every time the frequency is doubled.

2.4.10.2. *Variability*

The radiomobile environment fluctuates. The passage of vehicles and people, the wind in the trees, the opening of doors cause fluctuations in radioelectric paths and generate rapid variations in the observed signal. Combined with the movement of vehicles, these phenomena create variability in the propagation channel in space and time.

Signal variations are of a random nature. Statistical analyses will assess the impact of multi-paths on the transmission of radiomobile systems. The law of the field received and the average signal fading time, for example, are essential data to determine the size of the transmission equipment and the devices used to prevent it (interlacing, diversity).

Modeling of rapid variations has been proposed (Clarke 1968). Assuming that the mobile moves in an interference figure generated by the superposition of a large number of plane waves of independent amplitudes, phases and random directions, the received field (the complex envelope) is, by the application of the central limit theorem, a Gaussian variable. The envelope of the narrowband signal, the received power, then follows Rayleigh's law. This is the concept of Rayleigh fading.

The probability density of a random variable R following Rayleigh's law is given by the following equation:

$$f_R(x) = \frac{x}{\sigma^2} \exp\left(-\frac{x^2}{2\sigma^2}\right), \; x \geq 0$$
$$f_R(x) = 0, \; x > 0 \qquad [2.47]$$

It is characterized by the parameter σ where σ^2 represents the average power of the signal. It is observed in the presence of multiple paths in an obstructed link configuration where the contribution of each path is equivalent.

In the case of a predominant path (visibility, open environment such as suburbs, rural, etc.), the envelope of the narrowband signal, the received power, then follows Rice's theorem whose density of probability is given by:

$$f_R(x) = \frac{x}{\sigma^2} \exp\left(-\frac{x^2 + r^2}{2\sigma^2}\right) I_0\left(\frac{xr}{\sigma^2}\right), \; x \geq 0$$
$$f_R(x) = 0, \; x > 0 \qquad [2.48]$$

where I_0 is a modified Bessel function of the first kind and order 0.

Rician fading is characterized by the parameters r and σ. It is observed when one path is predominant. Rice's theorem corresponds to Rayleigh's law when $r = 0$ (absence of a direct path) and identifies a direct route and its preponderance. It is also characterized by the coefficient K (Rician parameter) defined by the equation:

$$K = 10 \log\left(\frac{r^2}{2\sigma^2}\right) \qquad [2.49]$$

The parameter K represents the ratio between the power of a direct path and the power contribution of secondary paths obeying Rayleigh's law. The larger parameter K, the greater the power of the direct path compared to multiple paths and the more the link is cleared. Conversely if the predominant path is weak (K≤ -5 dB), Rice's law is considered to be Rayleigh's law. The value of K depends on the environment (dense urban, suburban, rural, etc.).

Other laws, such as those of Weibull and Nakagami (Braun and Dersch 1991) also characterize the envelope of the mobile signal. They are less frequently used than Rice and Rayleigh's laws, which are more easily implemented in calculation software because they are generated by Gaussian variables. For more details concerning these various laws, characterizing the distribution of the rapid variations in radiomobile signals, the reader should refer to the recommendation ITU-R P.1057.

In order to determine the law of variation followed by the signal, statistical tests are applied, for example the Kolmogorov-Smirnov test (Barbot *et al.* 1992) on these rapid variations, resulting from the interferences of received waves.

Knowing the statistical characteristics of the signal (probability density, cumulative distribution function), it is possible to determine the parameters relevant to the operation of a radioelectric system, including:

– the probability of going below a certain level. When the signal strength is below the noise threshold tolerated by the receiver (thermal noise, interference, industrial noise, etc.), the signal is masked by noise. The receiver is then unable to correctly interpret the transmitted information;

– the statistical duration of fading. During the fading, the mass of information is lost. The slower the mobile moves, the longer the cut-off times.

Some orders of magnitude are given in Table 2.1 at 900 MHz (Guisnet 1998). We see that fading above 10 dB is very frequent: it appears in 10% of cases. For a pedestrian at 4 km/h, the average duration of signal fading is 40 ms; it is only 4 ms for a vehicle at 40 km/h. A high-speed mobile establishes communication with errors almost uniformly distributed in time, while a pedestrian sees this communication weakened or interrupted for a relatively long time.

		Average fading duration	
Level of signal fading with respect to average field	P(x < threshold) (%)	f = 900 MHz v = 4 km/h	f = 900 MHz v = 40 km/h
-5 dB	27%	80 ms	8 ms
-10 dB	10%	40 ms	4 ms
-20 dB	1%	12 ms	1 ms
-30 dB	0.1%	4 ms	0.4 ms

Table 2.1. *Statistical characterization of deep fade*

Variability due to movement can also be characterized by the Doppler spectrum. It is obtained from the power spectral density of the function $T(f_0,t)$. A pure frequency line at the transmission is enlarged in proportion to the speed of the mobile.

The Doppler spectrum is distributed over the frequency interval $[f_0 - f_d, f_0 + f_d]$ where f_0 is the emitted frequency, f_d is the maximum Doppler frequency given by the equation:

$$f_d = f_0 \frac{v}{c} \qquad [2.50]$$

where v is the velocity of the mobile and c the velocity of the electromagnetic wave.

The typical Doppler spectrum is given by the following equation:

$$S(f) = \frac{1}{\sqrt[2]{\left(f_d^2 - f^2\right)}} = \frac{1}{f_d \sqrt[2]{1 - \left(\frac{f}{f_d}\right)^2}} \qquad [2.51]$$

where f belongs to the interval $[- f_d, + f_d]$.

The Doppler spectra deduced from the measurements (Gollreiter 1994; Codit 1994; Dersch and Zollinger 1994) clearly show asymmetries in the spatial frequency domain that confirm, particularly in line-of-light situations, preferred directions of arrival.

The inverse of the width of the Doppler spectrum (the Doppler dispersion) is known as the coherence time. This is the time during which the function $T(f_0,t)$ can be considered to be almost constant: the time during which a moving receiver would not perceive the channel variability.

To combat deep fade, data interlacing and diversity techniques (microdiversity, frequency) are used (Fechtel and Meyr 1993; Visoz *et al.* 1998; Rappaport 2009).

2.4.10.3. *Frequency selectivity*

When the differences between the delays of the multiple paths are large, the transfer function is no longer constant over the entire width of the spectrum: the propagation loss depends on the chosen frequency. The channel is said to be frequency selective. Broadband channel modeling is therefore essential to assess the performance of a complete transmission chain, design new systems and ensure the quality of digital signal transmission.

Since multiple paths overlap in a linear fashion, the channel is generally represented by a time-variable linear filter of the WSSUS type (Wide Sense Stationary Uncorrelated Scattering) (Bello 1963) stationary in the broad sense where the diffusers are uncorrelated. The echoes perceived by the receiver come from a set of uncorrelated sources: two echoes with different propagation delays are uncorrelated. This property however is extremely difficult to verify but it is considered, in practice, to be applied over distances of a few wavelengths.

The radioelectric channel is represented by its impulse response that is variable over time, h(t,τ), with τ being the delay and t the dependence on time (and therefore in space too since the vehicle is moving). As a function with two variables, it expresses the three characteristics of the channel: the attenuation, the variability (t) and the selectivity (τ).

The dual variables, by Fourrier transform of τ and t, are respectively the frequency and the Doppler speed (Parsons 1992; Kattenbach and Fruchting 1995).

2.4.10.3.1. The different representations of the radiomobile channel

Four representations are thus possible, the connections of which can be represented in the following way (Figure 2.7) (Bello 1963); F and F^{-1} represent respectively the direct and inverse Fourier transforms.

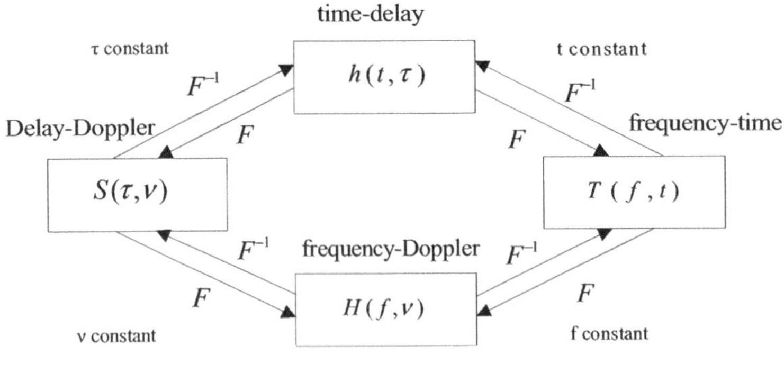

Figure 2.7. *Representation of the different Fourier transforms on impulse response*

Time-delay representation

The time-delay representation describes the output signal y(t) as a function of the input signal x(t) in the form of a convolution equation whose kernel is variable over time:

$$y(t)=\int_{0}^{\infty}h(t,\tau)x(t-\tau)d\tau \qquad [2.52]$$

where h(t,τ) is the response at instant t to a radioelectric impulse x(t) transmitted at t-τ. It defines the propagation channel and can differentiate the different echoes as a function of their propagation delays.

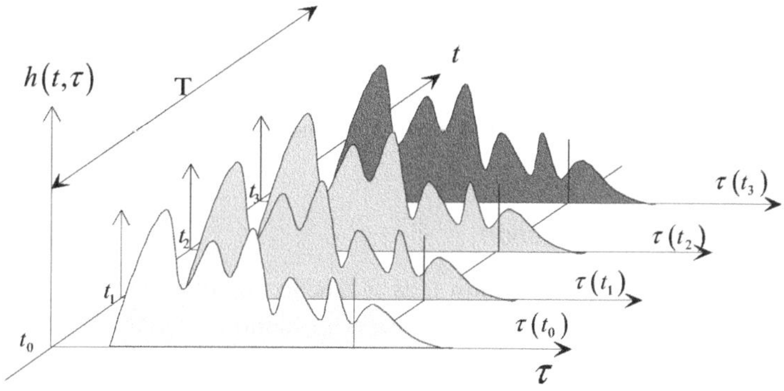

Figure 2.8. *Representation of the temporal evolution of the propagation channel impulse response*

The delay-Doppler-shift representation

The delay-Doppler shift representation S(τ,ν) is very attractive from a physical analysis point of view of propagation paths. It also allows you to follow the evolution of the different propagation paths when moving a mobile at constant speed. On the other hand, for a stable propagation path, the evolution of the frequency offset will provide information on its angle of arrival. This function S(τ,ν) is defined by:

$$y(t)=\int_{-\infty}^{+\infty}\int_{-\infty}^{\infty}x(t-\tau)S(\tau,\nu)e^{2\pi j\tau t}d\nu d\tau \qquad [2.53]$$

with:

$$h(t,\tau)=\int_{-\infty}^{+\infty} S(\tau,\nu)e^{2\pi j\nu t}d\nu \qquad [2.54]$$

Doppler frequency shift – frequency representation

The dual function H(f,ν), of function h(t,τ), represents the Doppler frequency shift as a function of frequency. This function is defined as follows:

$$Y(f)=\int_{-\infty}^{+\infty} X(f-\nu)H(f,\nu)d\nu \qquad [2.55]$$

where X(f) and Y(f) are respectively the Fourier transforms of x(t) and y(t).

This function directly identifies each frequency shift. It also characterizes the channel frequency selectivity.

Attenuation-time representation

The transfer function T(f,t) allows, as a function of time, multiple path effects (temporal attenuation, or when the mobile is in motion, spatial attenuation) to be studied and the characterization of the narrowband propagation channel. It is defined by the equation:

$$y(t)=\int_{-\infty}^{+\infty} X(f)T(f,t)e^{2\pi jft}df \qquad [2.56]$$

2.4.10.3.2. Broadband representation of the radiomobile channel

It is typical to qualify the selectivity of the channel by parameters deduced from the mean delay profile of the impulse response. The most used are the mean delay, delay spread, delay interval, delay window and channel coherence bandwidth (Failly 1989).

Average power density

The average power density P(τ) of the impulse response (average delay profile) is defined using h(t,τ) by the equation:

$$P(\tau)=\frac{1}{T}\int_0^T\left|h(t,\tau)\right|^2 dt \qquad [2.57]$$

It corresponds to an average, over a certain duration, where the WSSUS property applies. The duration T (Figure 2.9) is chosen so that over this time period, the measured impulse responses can be represented by a stationary and ergodic random process (Lavergnat and Sylvain 1997).

Figure 2.9 gives, in a microcellular context, an example of the variation of the impulse response |h(t,τ)| when the mobile turns a street corner. The received power is reduced by 30 dB and the shape of the response is modified in a significant way.

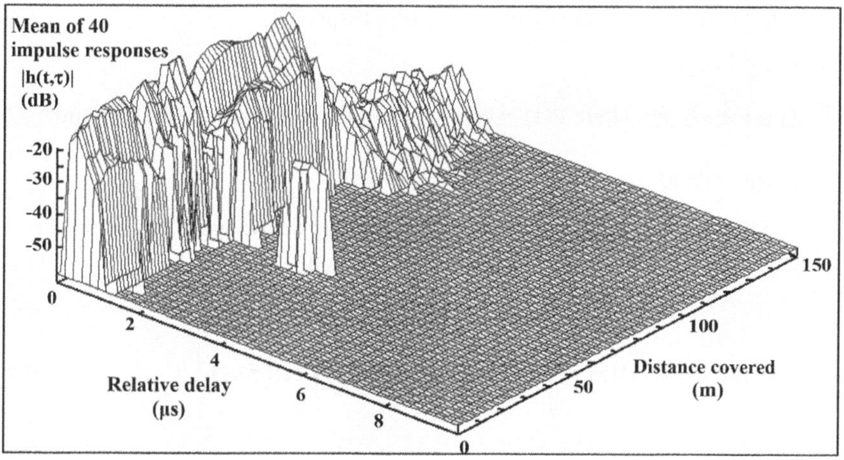

Figure 2.9. *Evolution of the impulse response: turning a street corner in the microcellular environment (Paris, 900 MHz, FTR&D sounder)*

Mean delay

The mean delay is the average of the delays weighted by their power. It is given by the first order moment of the impulse response:

$$\tau_m(t)=\frac{1}{P_m}\int_{\tau_{LOS}}^{\tau_3}(\tau-\tau_{LOS})P(t,\tau)d\tau \qquad [2.58]$$

where:

– τ_{LOS} is the propagation time in line-of-sight;

– τ_3 is the instant where P(τ) exceeds the cut-off threshold for the last time;

– P_m is the total energy of the impulse response, defined by the following equation:

$$P_m = \int_{\tau_0}^{\tau_3} P(\tau) d\tau \qquad [2.59]$$

where:

– P(t,τ) is the power density of the impulse response;

– τ is the excess delay;

– τ_0 is the instant where P(τ) exceeds the cut-off threshold for the first time.

The delay spread

The delay spread or the standard deviation of the delays weighted by their power is given by the second moment of the impulse response:

$$Delay-spread(t) = \sqrt{\frac{1}{P_m}\int_{\tau_0}^{\tau_3} \tau^2 P(\tau) d\tau - \left[\frac{1}{P_m}\int_{\tau_0}^{\tau_3} \tau P(\tau) d\tau\right]^2} \qquad [2.60]$$

The delay spread illustrates the risk of the occurrence of inter-symbol interferences and the disturbances that these strong distant echoes may induce.

Delay interval

The delay interval at X dB is defined as the time interval between instant τ_0 where the amplitude of the impulse response crosses a given threshold for the first time and instant τ_3 where this amplitude becomes lower than this given threshold for the first time.

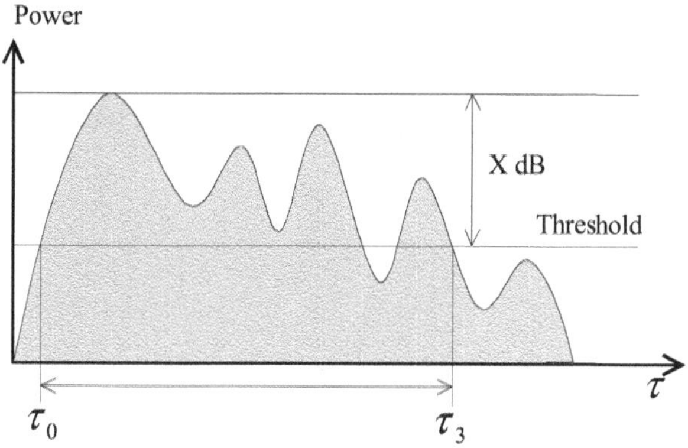

Figure 2.10. *Example of a power delay profile; highlighted by the delay interval at X dB*

Delay window

The delay window at y%, is the duration of the central portion ($\tau_2-\tau_1$) of the impulse response that contains y% of the total energy. Instants τ_1 and τ_2 are defined by the equation:

$$\int_{\tau_1}^{\tau_2} P(\tau)d\tau = \frac{y}{100}\int_{\tau_0}^{\tau_3} P(\tau)d\tau = \frac{y}{100}P_m \qquad [2.61]$$

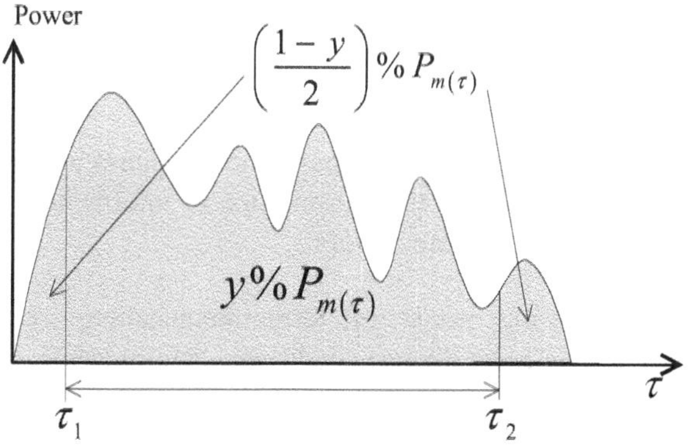

Figure 2.11. *Example of a power delay profile; highlighted by the delay window at y%*

Coherence bandwidth

The coherence bandwidth ($B_{coherence}$) of the channel is defined as follows. Here C(t,f) is the autocorrelation of the transfer function (Fourier transform of the impulse response power):

$$C(t,f)=\int_{\tau_0}^{\tau_3} P(\tau)e^{-2\pi jft}d\tau \qquad [2.62]$$

The correlation bandwidth is defined as the frequency for which |C(t,f)| is equal to x% of C(t,f = 0) (Parsons 1992). This indicates the selective attenuation amplitude as a function of frequency separation. The correlation bandwidth is therefore the frequency from which the autocorrelation function of the transfer function crosses a given threshold.

To analyze the experimental data, the ITU-R recommends the use of delay intervals for thresholds of 9, 12 and 15 dB below the peak value and the delay windows for 50%, 75% and 90% energy and a correlation bandwidth for a 50% and 90% correlation.

The correlation band is linked to the delay spread by the following equation (Lee 1986):

$$Bcoherence.Delay-spread=\frac{1}{2\pi} \qquad [2.63]$$

2.5. Propagation modeling

The development of telecommunications services requires a good knowledge of the propagation of waves ensuring the transmission of information, whether this be sound, electromagnetic, radioelectric or light.

The constantly growing needs of telecommunications require having a propagation model in ever more diverse conditions: environments, frequency range, bandwidth, etc.

The total propagation loss (A) includes, in addition to the free space loss (A_0), an additional loss A_s which reflects the influence of many environmental factors causing loss (gas, hydrometeors, etc.), mask effect due to obstacles (buildings, walls, partitions, furniture, people, vegetation, etc.) or variability due to interference. In practice, it is also necessary to consider the losses in the feeder lines of the antennas, and more generally of all the devices which are inserted between the transmitters or receivers and the antennas.

There are different types of models: deterministic, empirical and semi-empirical models.

Deterministic models rely on the fundamental laws of physics. They serve as reference models. The computing time is relatively high, however.

Empirical models are based on the analysis of a large number of experimental measurements according to different parameters such as the frequency, distance and height of the antennas. They are robust, fast and do not require geographic databases. They are suitable for sizing systems but fairly imprecise at short distances.

Semi-empirical models combine an analytical formulation of physical phenomena (reflection, transmission, diffraction, scattering) and statistical adjustment using experimental measurements. They are fast, precise and robust. On the other hand, they require consideration of the environment (troposphere, geographic databases – topography, land use, morphology, building contours, street axis, etc.).

We will develop below tropospheric propagation, both line-of-sight and in non-line-of-sight, the propagation of radioelectric waves in rural, suburban and urban areas, the propagation of radioelectric waves inside buildings and the propagation of radioelectric waves in broadband and ultra-wideband contexts. There are many applications: fixed links (FH, WIMAX, WIFI, etc.), mobile links (GSM, UMTS, 3G, 4G, 5G, etc.), ULB links (UWB) very high speed short range etc.

2.5.1. *Tropospheric propagation*

The Earth's atmosphere is the gaseous envelope which surrounds the Earth, and which participates with it in its different movements. The troposphere is the lowest atmospheric layer. It is characterized by a regular decrease in temperature depending on the altitude, approximately -5 to -6°C on average every 1,000 meters. It is in the troposphere that most weather phenomena occur, including the formation of clouds. The upper limit of the troposphere ranges from 8 kilometers (at the pole) to

18 kilometers above sea level (at the equator) depending on geographic latitude, seasons and weather conditions. This is the tropopause where the temperature varies from 190 K at the equator to 220 K at the poles. It slows upward convection movements and constitutes an upper limit for clouds (except cumulonimbus).

2.5.1.1. *Line-of-sight radioelectric propagation*

The presence of the Earth and the atmosphere brings into play different physical mechanisms (reflection, refraction, diffraction, diffusion) which modify the propagation conditions and influence the level of the received field, even if both ends of the link are in radioelectric line-of-sight. After defining the notion of radioelectric visibility, we will examine the different physical mechanisms encountered (attenuation (gas, hydrometeors – rain, fog, cloud, hail), reflection on the ground, refraction in the atmosphere, diffraction, etc.).

A link is considered to be in radioelectric line-of-sight if the first Fresnel ellipsoid delimiting the space domain where almost all of the energy passes are not engaged. Diffraction phenomena by possible obstacles located beyond the first Fresnel ellipsoid therefore have a negligible influence on the level received. With the radius of the ellipsoid being inversely proportional to the frequency, it is necessary to raise the antennas more at lower frequencies.

2.5.1.1.1. Attenuation in the atmosphere

Attenuation by gases

Attenuation by gases results from the molecular resonance of oxygen and water vapor.

The oxygen molecule has a permanent magnetic moment. Its coupling with the magnetic field of an incident electromagnetic wave causes absorption by resonance at certain frequencies. Around 60 GHz corresponds, in particular, to a coupling between the intrinsic moment of the electron (spin) and the rotational energy of the molecule, which generates a series of absorption lines fairly close to each other in the spectrum.

The water vapor molecule behaves like an electric dipole. Its interaction with an incident wave disorients the molecule by generating additional internal potential energy. The maximum attenuation reached around 22 GHz results from the resonance of the water molecule; it begins to rotate on itself absorbing a large proportion of the incident electromagnetic energy.

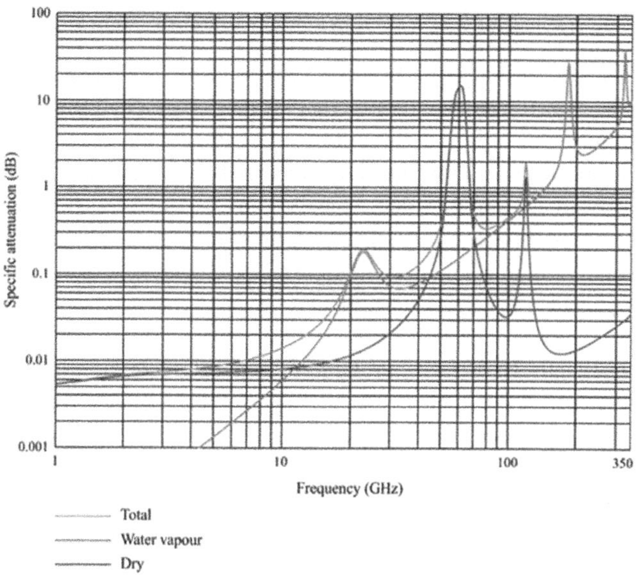

Figure 2.12. *Specific attenuation (dB/km) due to atmospheric gases (O_2 and H_2O) and total (ITU-R P.676)*

Figure 2.12 shows the dry air (Dry), water vapor only with a density of 7.5 g/m^3 (water vapor), and total (Total) specific attenuation from 1 to 350 GHz at sea-level for the mean annual global reference atmosphere given ITU-R P.676.

A very accurate way to estimate the attenuation by gases is to consider the contribution of all the absorption lines of oxygen and water vapor and the continuous absorption spectrum linked to water and ice. Different models exist in the literature ITU-R P.676 (Liebe 1983, 1985; Salonen 1990).

The reference model is that of Liebe, the MPM93 model (Liebe 1983). It calculates the refractive index linked to oxygen and water vapor in the atmosphere, as well as the attenuation linked to each of these components, for frequencies up to 1,000 GHz (Liebe 1983, 1985, 1989 and 1993). The input parameters of this model are pressure, temperature, relative humidity measured on a vertical profile of the Earth's atmosphere, as well as the frequency (COST 255 1999).

Numerical application: the linear attenuation due to atmospheric gases is, for an average atmosphere (7.5 g/m^3), approximately 0.2 dB/km and 15 dB/km respectively at 20 and 60 GHz.

Attenuation by hydrometeors

Attenuation due to rain is particularly related to the knowledge of precipitation intensities. The specific attenuation γ_r (dB/km) is obtained from the precipitation intensity R (mm/h), according to the following mathematical equation: $\gamma_r = kR^{\alpha}$. The coefficients k and α depend on the frequency and polarization (see ITU-R P.837). For an intensity of 20 mm/h (value exceeded for 0.1% of the time in Belfort, for example), the orders of magnitude of the attenuations due to rain are 2 and 8 dB/km respectively at 20 and 60 GHz. For an intensity of 40 mm/h, value exceeded for 0.01 of time, the values are 3.2 and 13 dB/km.

Recommendation ITU-R P.837 provides a model for determining the rain intensity R_p, which is exceeded for a given percentage p of the average year and at a given location, as well as examples of global rain distribution maps (mm/h) exceeded for 0.01% of the average year. The data files were established on the basis of data collected over fifteen years by the European Centre for Medium-Range Weather Forecasts (ECMWF). Figure 2.13 shows the change in specific loss (dB/km) due to rain.

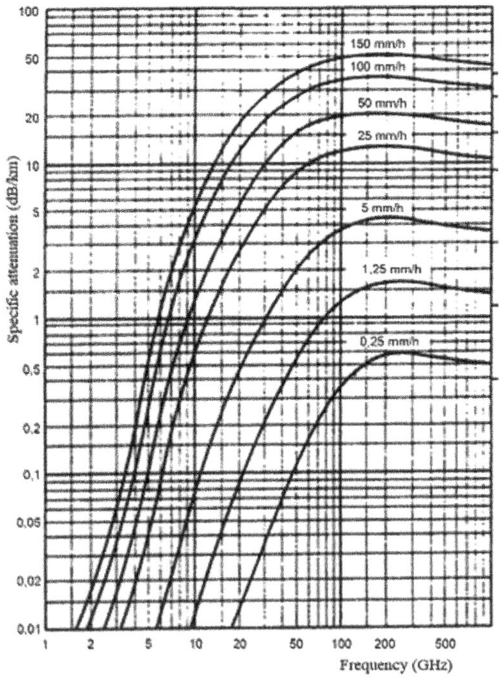

Figure 2.13. *Specific attenuation (dB/km) due to rain as a function of the frequency (ITU-R P.837)*

A number of models exist in the literature. The most cited names are: ITU-R, Bryant, Crane, DAH (Dissanayake-Alnutt-Haidara), Garcia, Karasawa, Leitao-Watson, Matricciani, SAM, Sviatogor, Assis-Einluft, Misme-Waldteufel, etc. (COST 255 1999).

Reflection on the ground

Reflection occurs when the wave encounters a surface with large dimensions (Ney 2004) and has small irregularities compared to the wavelength. The reflected field is linked to the incident field via Fresnel equations (Ney 2004). A distinction is made between specular reflection and diffuse reflection (Figures 2.14a and 2.14b).

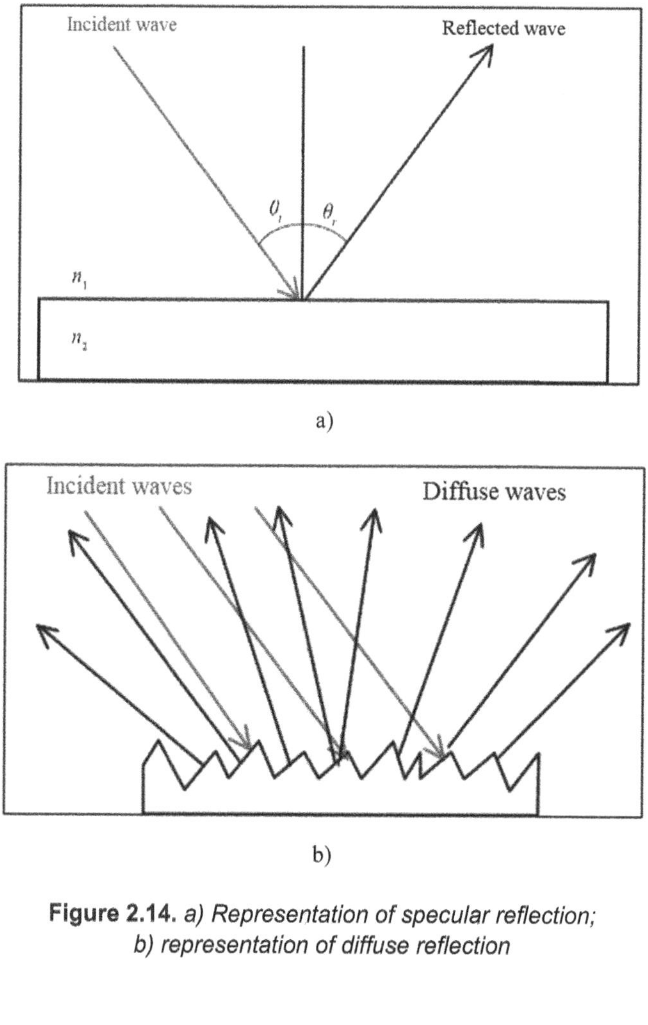

Figure 2.14. *a) Representation of specular reflection; b) representation of diffuse reflection*

Specular reflection

Specular reflection, a phenomenon common to all frequencies, is due to a perfectly flat homogeneous surface. It is caused by obstacles such as the ground, building facades and flat surfaces. The propagation attenuation induced by such reflections follows from Fresnel relations and depends on the dielectrics of the reflecting surface (conductivity σ, relative permittivity ε_r or its complex relative permittivity $\varepsilon'_r = \varepsilon_r - j60\sigma\lambda$). The relative magnetic permeability is always close to 1.

Figure 2.15 gives an example of the variation of the coefficients (real part and modulus) of reflection and transmission in vertical (hard) and horizontal (soft) polarization of wet soil at 1 GHz.

Wet soil at this frequency is characterized by a relative permittivity equal to 30 and a conductivity equal to 0.18 (S/m).

On the ground at low incidence, that is to say for very small angles of inclination (angle of incidence close to 90°), the reflection factor is always close to -1; the reflection occurs with phase inversion whatever the polarization (Boithias 1988). In horizontal polarization, the reflection factor remains close to -1 for fairly large angles of inclination. In vertical polarization, on the contrary, it decreases to a minimum, which is smaller when the frequency is high (Brewster incidence), then increases beyond (Figure 2.15b). When the angle of inclination reaches 90° (normal incidence), the two polarizations are equivalent (Boithias 1988).

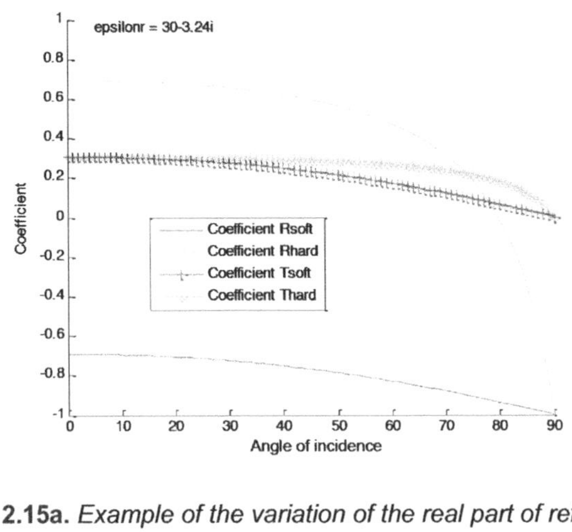

Figure 2.15a. *Example of the variation of the real part of reflection and transmission coefficients of wet soil at 1 GHz in vertical (hard) and horizontal (soft) polarizations*

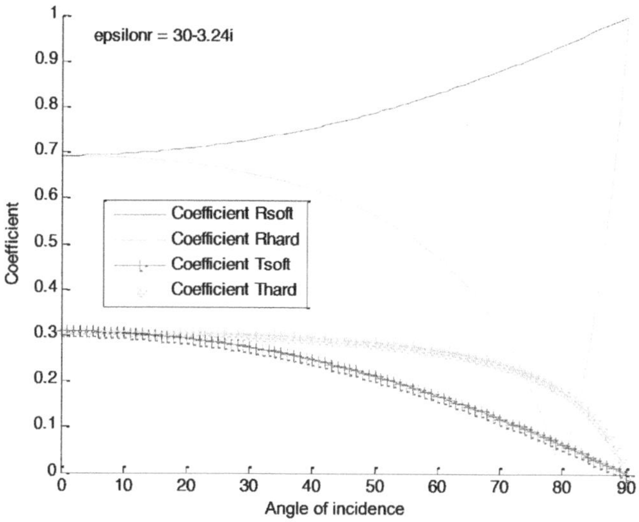

Figure 2.15b. *Example of the variation of the modulus of the reflection and transmission coefficients of wet soil at 1 GHz in vertical (hard) and horizontal (soft) polarizations*

Diffuse reflection

Diffuse reflection is due to reflections from surfaces that are not flat but rough; surfaces with uneven height at different points. As a result, an incident wave is no longer reflected in a single direction but is broadcast in multiple directions. In order to specify whether a reflection is diffuse or specular, the Rayleigh criterion is generally used, namely the consideration of the height of the surface irregularities (H) and the angle of inclination (i). A height H irregularity will then create, for two waves reflected by the surface, at most a difference in path $\delta = 2H\sin(i)$ and therefore a phase difference $\Delta\Phi = 4\pi H\sin(i)/\lambda$ (Figure 2.16).

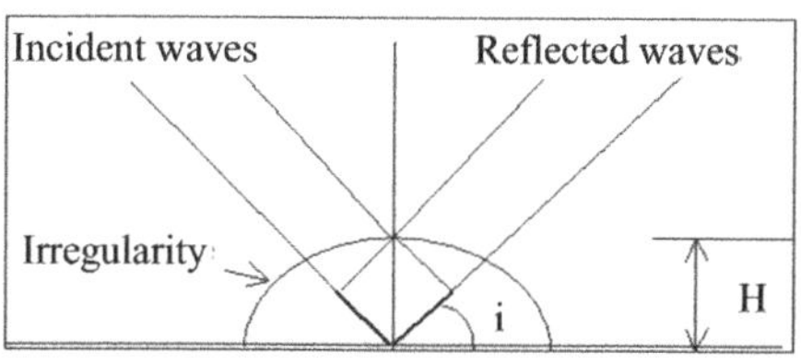

Figure 2.16. *Difference in path created by a surface irregularity with height H*

When the height H is small enough for these two waves to be in phase, this concerns the previous case of specular reflection. Otherwise the surface is considered to be rough. In summary, the reflection is diffuse, according to Rayleigh's criterion, when $\Delta\Phi > \pi/2$ so when $H > \lambda/8\sin(i)$. The roughness therefore depends on the frequency, the angle of incidence and the height of the irregularities.

It is possible to calculate the power reflected by a rough surface by multiplying the specular reflection coefficient for each of the horizontal H and vertical V polarizations by a diffusion coefficient ρ:

$$R_{V,H}^{\mathrm{mod}} = R_{V,H} \cdot \rho \qquad [2.64]$$

with:

$$\rho = e^{-8\pi^2 \left(\frac{\sigma_h}{\lambda}\right)^2 \cos^2\theta} \qquad [2.65]$$

where:

– σ_h is the standard deviation of the height distribution of the irregularities;

– θ is the angle of incidence with respect to the normal.

The modeling of the reflection on the ground (two rays)

In communications systems (fixed or mobile connections), the distance between the transmitter and the receiver is generally a few tens of kilometers. The earth is therefore assumed to be flat.

The two-ray propagation model based on geometric optics (Figure 2.17) considers the combination of the direct ray and the ray reflected on the ground. The received signal results from the interference of two signals having traveled different paths. Depending on the relative phase of these, the field received can be maximum or minimum.

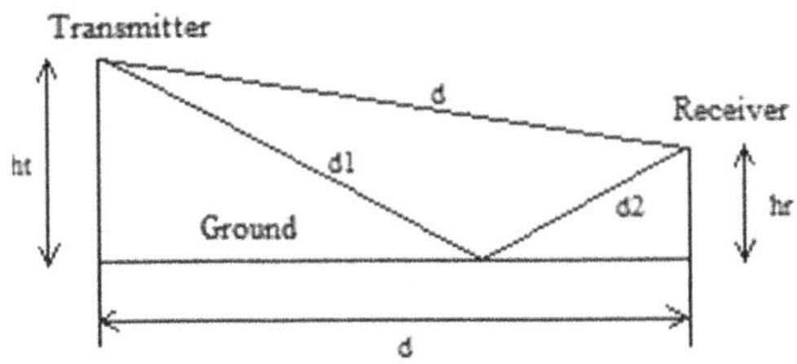

Figure 2.17. *Two-line model*

The received field (E_{total}) is the sum of the direct ray and the ray reflected on the ground:

$$E_{total}(d) = E_d(d).\left(1 + R.e^{-j\Delta\varphi}\right) \qquad [2.66]$$

where:

– $E_d(d)$ is the free-space field at the receiver;

– R is the reflection coefficient;

– $\Delta\varphi$ is the phase difference between both paths.

Transmitting and receiving antennas are assumed to be not very directive.

If h_t and h_r are respectively the heights of the transmitting and receiving antennas and d the distance between the transmitter and the receiver, the path difference in between the two rays, based on geometric considerations is according to the image method:

$$\Delta d = d_1 - (d_2 + d_3) \qquad [2.67]$$

so:

$$\Delta d = \sqrt{(h_t + h_r)^2 + d^2} - \sqrt{(h_t - h_r)^2 + d^2} \qquad [2.68]$$

Since the distance d is large with respect to the antennas, the path difference is written according to the Taylor series approximation:

$$\Delta d = \frac{2 \times h_t \times h_r}{d} \qquad [2.69]$$

whereby:

$$\Delta\varphi = \frac{2\pi}{\lambda}\Delta d = \frac{4\pi}{\lambda}\frac{h_t h_r}{d} \qquad [2.70]$$

For large values of d ($d \gg \sqrt{h_t h_r}$), and with R # -1, $\Delta\phi$ is small, giving:

$$E_{total}(d) \approx E_d(d)\Delta\varphi = E_d(d)\frac{4\pi}{\lambda}\frac{h_t h_r}{d} \qquad [2.71]$$

The received power in free space was:

$$P_r(d) = G_t G_r \frac{\lambda^2}{(4\pi d)^2} P_t \qquad [2.72]$$

This therefore becomes:

$$P_r(d) = \left(\frac{4\pi}{\lambda}\frac{h_t h_r}{d}\right)^2 G_t G_r \frac{\lambda^2}{(4\pi d)^2} P_t = \frac{h_t h_r}{d^4} G_t G_r P_t \qquad [2.73]$$

The power received is thus, at long distance, inversely proportional to d^4, that is a decrease of 40 dB per decade, a much faster decrease than that received in free space. It will also be noted that the power received and the attenuation become independent of the frequency.

The attenuation, expressed in dB, is therefore:

$$\begin{aligned} Aff &= -10\log_{10} G_t - 10\log_{10} G_r - 20\log_{10} h_t - \ldots \\ &\ldots 20\log_{10} h_r + 40\log_{10} d \end{aligned} \qquad [2.74]$$

This situation, where the direct ray and the reflected ray have slightly different paths, must therefore be avoided. In order to overcome the interference between the two signals, it is a good idea to place the receiving antenna so that the reflected ray is masked by an obstacle (Figure 2.18).

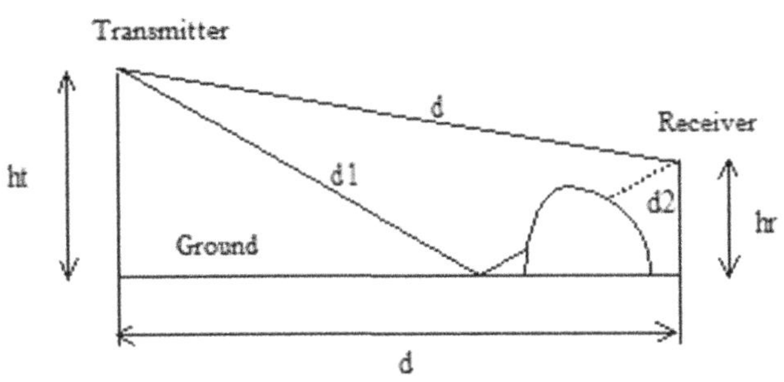

Figure 2.18. *Diagram showing the blocking of the reflected ray with an obstacle*

The amplitude of the field received depends on the roughness of the soil. For a maritime connection, it is advisable to limit the effect of the sea which has a strong reflectance, when forming the connection, where possible, so that the reflecting point is located on an island which has a weaker reflectance (Figure 2.19).

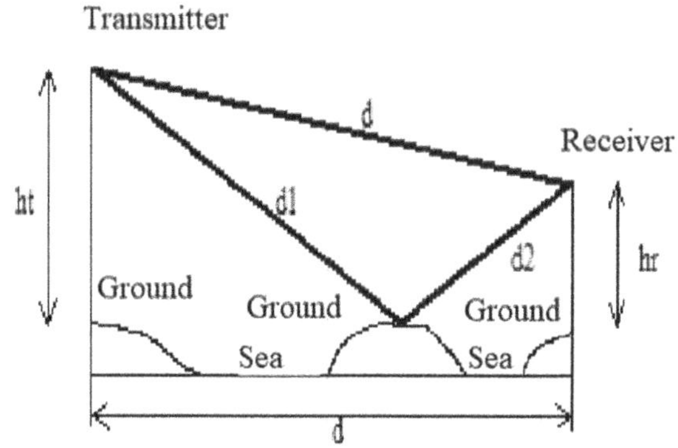

Figure 2.19. *Diagram showing the path reflected on an island to limit the effect of the reflected path on a maritime link*

Refraction in the atmosphere

The refractive index

In radioelectricity, the troposphere is considered to be a dielectric with a refractive index close to 1, but whose variations, although small, play an important role.

The refractive index n is given by:

$$n = \sqrt{\varepsilon_r \mu_r} \qquad [2.75]$$

where:

– ε_r is the relative dielectric constant;

– μ_r is the relative permeability.

The average value of the refractive index n, at ground level, varies around 1.0003. Since variations only appear at the fifth and sixth decimals, we use the refractivity N which gives a practical value of the refractive index n. These two indices are linked by the following equation (ITU-R P.834):

$$N = (n-1)10^6 \qquad [2.76]$$

The refractivity can be obtained either by measuring it directly using a refractometer, or by calculating it from meteorological data using the following equation:

$$N = 77.6\frac{P}{T} + 3.73 \times 10^5 \frac{e}{T^2} \quad [2.77]$$

where:

– T is the absolute temperature in K;

– P is the atmospheric pressure in hectopascals (or mb);

– e is the partial pressure of water vapor in hectopascals, with:

$$e = \frac{He_s}{100} \quad [2.78]$$

where:

– H is the relative humidity of air (quantity of water vapor contained in the air as a percentage compared to the maximum quantity that this air could contain at the same temperature);

– e_s is the saturation vapor pressure. It is linked to the air temperature t (° C) by the following expression (ITU-R P.453):

$$e_s = a \exp\left(\frac{bt}{t+c}\right) \quad [2.79]$$

The values of coefficients a, b and c for liquid water are a = 6.1121, b = 17.502 and c = 240.97. These are applicable between – 20° C and + 50° C with an accuracy of +/– 0.2%.

The first term for the expression of N as a function of pressure and temperature is the dry component; the second term, depending on humidity and temperature, is the wet component. The dry component is the strongest component, contributing 60 to 80% of the index value. Graphs give the value of the dry and wet components according to the different meteorological parameters (Boithias 1983). In summer, the wet component tends to increase and the dry component tends to decrease. In winter, the wet component is weak while the dry component tends to increase due to the low temperatures.

As pressure, temperature and humidity vary with altitude, the air index also varies with distance and altitude. These variations depend essentially on climatic phenomena such as subsidence, night-time radiation, or even the nature of the soil (presence of water on the soil). The variability with the distance is particularly strong on mixed paths (land-sea) where islands are traversed. Vertical variability, generally the only consideration, is very important in propagation.

Radioelectric wave trajectories

The path along which radioelectric energy travels is called a ray. The propagation of waves in the troposphere is mainly a function of the value of the refractive index and the index gradient. In a continuous medium (a medium where the refractive index n(h) varies continuously with altitude), this law is written (Descartes law):

$$n(h)\cos(\varphi(h)) = Cste \quad [2.80]$$

where:

– $\varphi(h)$ is the angle between the ray and the horizontal, at altitude h;

– Cste is a constant;

with:

$$\varphi(h) = \frac{\pi}{2} - \theta(h) \text{ (Figure 2.20)} \quad [2.81]$$

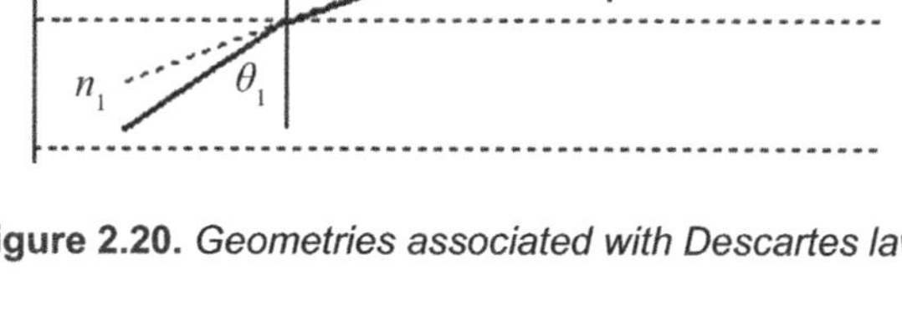

Figure 2.20. *Geometries associated with Descartes law*

When the stratification is spherical, it is necessary to modify this relation. The application of Descartes law to a medium with spherical geometry is called Bouguer's law. For a continuous medium, we have the following (Boithias 1983):

$$n(h)r(h)\cos(\varphi(h)) = Cste \quad [2.82]$$

where:

– $\varphi(h)$ is the angle between the ray and the local horizontal at altitude h;

– r(h) is equal to (a + h) where a is an arbitrary reference taken with respect to the spherical center of symmetry. For example, if the center of the Earth is taken as the center of symmetry, a will have the value of the Earth's radius (approximately 6,370 km).

The radius of curvature

A radioelectric ray crossing the troposphere undergoes curvatures due to the vertical gradient of the refractive index. The radius of curvature at a point in space is then contained in the vertical plane and is expressed in the form (ITU-R P.453):

$$\frac{1}{\rho} = -\frac{\cos\varphi}{n}\frac{dn}{dh} \quad [2.83]$$

where φ is the angle of the ray trajectory with the horizontal.

The radius of curvature is defined as being positive when its concavity is oriented towards the surface of the Earth. This phenomenon practically does not depend on frequency, if the gradient of the index does not vary significantly over a distance equal to the wavelength.

If the antennas are near the ground, we have cos φ = 1 and since n = 1, we can write:

$$\frac{1}{\rho} = -\frac{dn}{dh} = \frac{1}{10^6}\frac{dN}{dh} \quad [2.84]$$

We then see that the path curvature is proportional to the refractivity gradient and if it is constant, the trajectories are arcs of a circle.

In particular, for a standard atmosphere characterized by a vertical gradient, equal to - 39 N units/km (normal refractivity gradient), the radius of curvature is

25,640 km, or about four times the radius of curvature of the Earth. It is a conventional value of the vertical coefficient gradient used for refraction studies. It corresponds approximately to the median value of the gradient over the first kilometer altitude in temperate regions.

If the algebraic value of the coefficient gradient is greater than – 39 N units/km, we say that there is infra-refraction. In particular if the gradient is zero, the atmosphere is linear, the paths are straight lines.

If the algebraic value of the coefficient gradient is less than – 39 N units/km, we say that there is super-refraction. The paths are more curved.

For a gradient of – 157 N units/km, the radius of curvature of the paths is equal to that of the Earth, and the radius is therefore parallel to the surface of the Earth.

If, however, the gradient is less than – 157 N units/km, the curvature of the paths is greater than the terrestrial curvature. The rays return to the ground where they are generally reflected. This is known as duct propagation. The energy of radioelectric waves remains confined and propagates with a much lower attenuation than in a homogeneous atmosphere. This is guided propagation (Figure 2.21).

There are two main types of ducts: surface ducts (the lower limit is on the ground) and high ducts (ITU-R P.453). They are characterized by their thickness (difference in altitude between the limit, lower and upper surfaces), their height (height above the ground of the lower surface of the higher duct) and intensity (difference between the maximum and minimum values of the modulus of the refractive index).

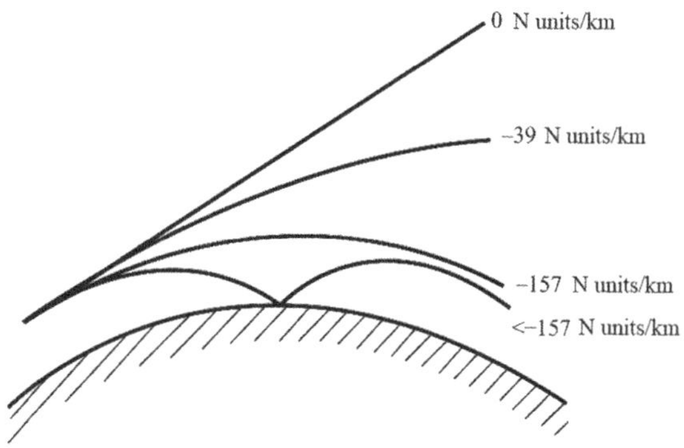

Figure 2.21. *Paths of radioelectric waves as a function of refractivity gradient*

Particular meteorological or geographical conditions lead to refractivity gradient values that deviate notably from the standard value. Thus, temporary temperature inversions in the atmosphere cause the radioelectric beams to deflect upwards and thus lead to communication disruptions. These phenomena sometimes appear in winter on television networks.

In order to account for refraction, numerical methods of ray tracing in the Earth's atmosphere generally use an equivalent Earth ray. This is the radius of a fictitious, spherical and atmosphere-free earth so that the radioelectric paths are rectilinear. The altitudes and the distances along the ground are the same as on the real Earth surrounded by an atmosphere whose vertical refractivity gradient is constant. For an atmosphere with a normal refractivity gradient, the equivalent terrestrial radius is approximately equal to 4/3 of the real radius of the Earth, which corresponds to approximately 8,500 km. The multiplicative factor of the terrestrial radius (ratio of the terrestrial radius equivalent to the real radius of the Earth) is linked to the vertical gradient dn/dh of the refractive index and to the terrestrial radius of the Earth a by the relation:

$$k = \frac{1}{1 + a\dfrac{dn}{dh}} \quad [2.85]$$

Numerical methods (parabolic equation methods, integral equation methods, etc.) allow the electromagnetic field to be vertically assessed at any point according to the profiles of the refractive indices (Levy 2000).

2.5.1.2. *Non-line-of-sight radioelectric propagation*

A link is considered to be non-line-of-sight radioelectric if the first Fresnel ellipsoid delimiting the space domain, where almost all of the energy passes, is far from being released. Diffraction phenomena by possible obstacles, located in the vicinity of the link, therefore have a significant influence on the level received. These conditions are always met when one end of the link is beyond the horizon of the other. They can also be carried out below the horizon when the dimensions of the Fresnel ellipsoid are large, particularly in decametric waves and in longer waves.

The physical mechanisms allowing electromagnetic waves to bypass obstacles formed by the relief and the curvature of the Earth are:

– diffraction around the supposedly spherical surface of the Earth;

– diffraction by the vertex of relatively thin edges;

– diffusion by heterogeneities;

– the reflection on the upper layers;

– duct propagation.

2.5.1.2.1. Diffraction around the Earth's surface

Wave diffraction around the Earth's sphere is one of the earliest known propagation difficulties and is known as ground wave propagation. Following the work of mathematicians such as Poincaré, Sommerfeld, Van der Pol, Bremmer, etc., the general expression of the radio-electric field is given by the following equation (Boithias 1998):

$$\frac{E}{E_0} = \sum_{n=1}^{n=\infty} A_n(d)\, g_n(h_1)\, g_n(h_2) \qquad [2.86]$$

where:

– E is the field at point;

– E_0 is the field in free space at the same distance;

– d is the distance between the transmitter and the receiver measured along the large circle;

– h_1 and h_2 are the heights of the antennas above the terrestrial sphere;

– A_n (d) and g_n (h) are complex functions of the distance and height of each antenna, respectively. They also depend on the characteristics of the soil, the frequency and the polarization.

2.5.1.2.2. Diffraction by the vertex of relatively thin edges

Diffraction occurs when a wave meets the edge of an obstacle (mountain, hill, buildings, etc.) whose dimensions are large with respect to wavelength. It is one of the most important factors intervening in the propagation of radioelectric waves.

The use of geometric theory of diffraction (GTD) (McNamara *et al.* 1990) makes it possible to represent this phenomenon in the form of rays. The theory shows that the additional attenuation (besides free space) of diffraction (given by GTD or uniform theory of diffraction (UTD)) has a frequency dependence in 10 log (f) with regard to diffraction by an edge. It is present for the propagation of waves in the frequency bands up to a few GHz. Beyond this, and especially for frequencies above 15 GHz, the diffraction loss can become considerable compared to a simple reflection loss.

To be able to evaluate an order of magnitude of the diffraction attenuation in the frequency bands that interest us, let us consider an edge without thickness at the top at the height h (positive or negative) compared to the line that joins the transmitter to the receiver. Here, d is the total distance and d_1 and d_2 are the respective distances from the edge to the transmitter and to the receiver (Figure 2.22).

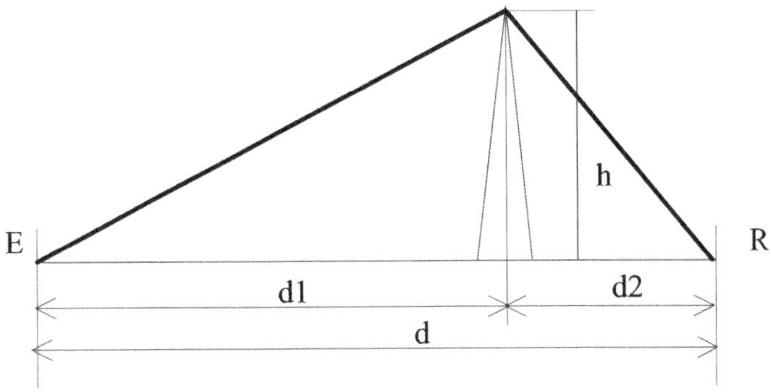

Figure 2.22. *Representation of a sharp diffracting edge*

If we take as a variable:

$$\nu = h\sqrt{\frac{2}{\lambda}\left(\frac{1}{d_1}+\frac{1}{d_2}\right)} \qquad [2.87]$$

the field at the receiver is given in amplitude and in phase, relative to the free space, by the following expression:

$$\frac{E}{E_0} = \frac{1}{1+j}\int_{\upsilon}^{\infty} e^{j\pi\frac{t^2}{2}}\,dt \qquad [2.88]$$

where E_0 is the field existing in the absence of an edge.

The ratio of the corresponding powers is then written:

$$\frac{P}{P_0} = \frac{1}{2}\left[\left(\frac{1}{2}-\xi(\nu)\right)^2+\left(\frac{1}{2}-\eta(\nu)\right)^2\right] \qquad [2.89]$$

with $\xi(\nu)$ and $\mu(\nu)$ being Fresnel integrals:

$$\xi(v) = \int_0^v \cos\frac{\pi t^2}{2} dt$$

$$\mu(v) = \int_0^v \sin\frac{\pi t^2}{2} dt$$

If v is negative, that is when the vertex of the edge is below the line connecting the transmitter to the receiver, P/P_0 tends to oscillate towards the free space level, whereas for positive v, P/P_0 decreases regularly as the obstruction of the edge increases. For zero v, the transmitter and the receiver are aligned with the top of the edge and the attenuation is 6 dB (in optics we have visibility while in radio we lose 6 dB).

In the case of the obstruction, we have approximate expressions (Boithias 1983):

$$10\log\frac{P}{P_0} = -13 - 20\log v \qquad [2.90]$$

equation valid particularly for $\nu > 1.5$:

$$10\log\frac{P}{P_0} = -6.9 + 20\log\left[\sqrt{(v-0.1)^2+1} - v + 0.1\right] \qquad [2.91]$$

more particularly valid for $\nu > -0{,}7$ and usable in the vicinity of $v = 0$.

Figure 2.23 gives the level of attenuation as a function of the height of the edge, relative to the transmitter-receiver axis.

In the literature, there are several other models that describe diffraction phenomena. The most common models are those of Bullington, Longley-Rice, Deygout, etc. These are models that have been greatly inspired by the work of Van Der Pol, Bremmer and Norton. In addition, these models are continuously evolving in order to get as close as possible to reality (model with several edges, etc.).

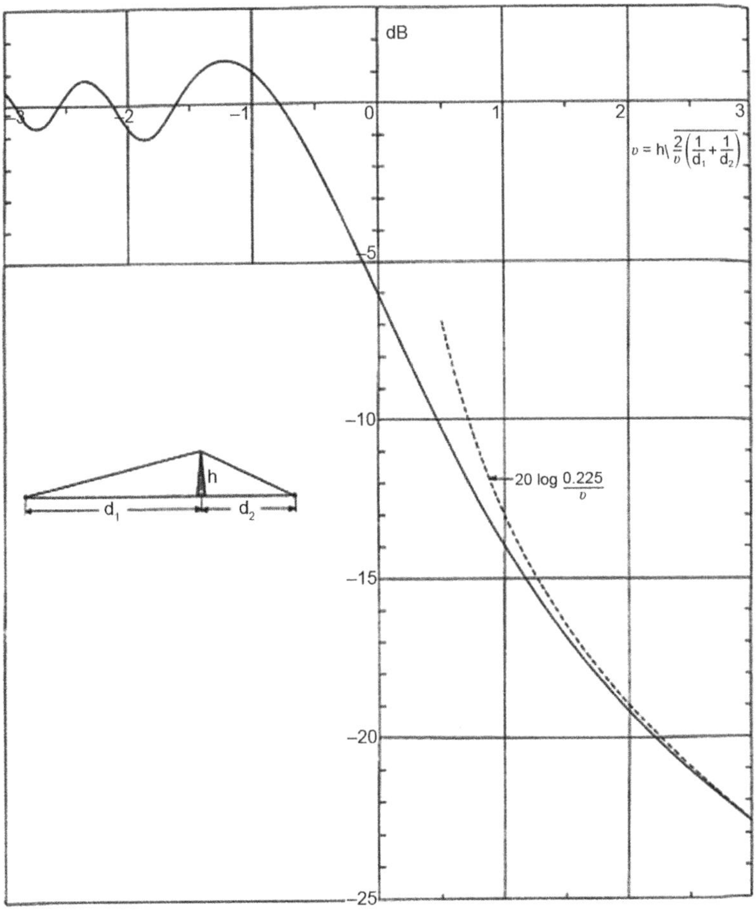

Figure 2.23. *Attenuation due to diffraction off an edge*

2.5.1.2.3. Diffraction due to heterogeneities

The real atmosphere is not homogeneous. It is in perpetual motion. Its refractive index presents significant spatiotemporal variations. In fact, it can undergo very rapid fluctuations and small amplitudes at each point, linked to the swirling movements of air particles. It may also contain non-gaseous particles such as raindrops, hailstones, dust, etc. These different heterogeneities present in the Earth's atmosphere will allow non-line-of-sight propagation of electromagnetic waves by diffusion. We distinguish more particularly the tropospheric diffusion and the diffusion by rain.

Tropospheric scattering

Tropospheric diffusion is linked to fluctuations in the refractive index. The propagation mechanism of an electromagnetic wave by tropospheric scattering is the following (Figure 2.24). The heterogeneities of the refractive index located in the volume common to the antenna beams, receiving energy from the transmitting antenna, return a small part of it in all directions and in particular towards the receiving antenna. As these heterogeneities fluctuate over time, the level of energy received undergoes the same fluctuations. The study of the energy received therefore has two stages, the study of the average level received on the one hand, the study of fluctuations about this average level on the other.

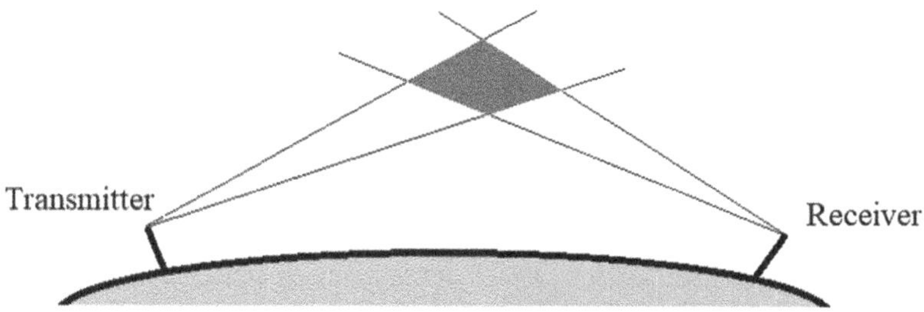

Figure 2.24. *Propagation of an electromagnetic wave by tropospheric scattering*

The average level received is mainly related to the average values of the index gradient in the volume common to the antenna beams. It is given by the following empirical relation:

$$A = 30\log_{10}(f) + 30\log_{10}(d) + 1.5G_c + 102 \qquad [2.92]$$

where:

– A is the attenuation between isotropic antennas;

– d is the distance in km;

– f is the frequency in MHz;

– G_c is the gradient of the refractive index in a common volume (N units/km). The value of G_c is generally negative.

The rapid fluctuations respect Rayleigh's law.

Figure 2.25 gives an example of tropospheric scattering observed on a 105 km link on the plain of Alsace at 468, 915 and 2,205 MHz (Blanchard 1999).

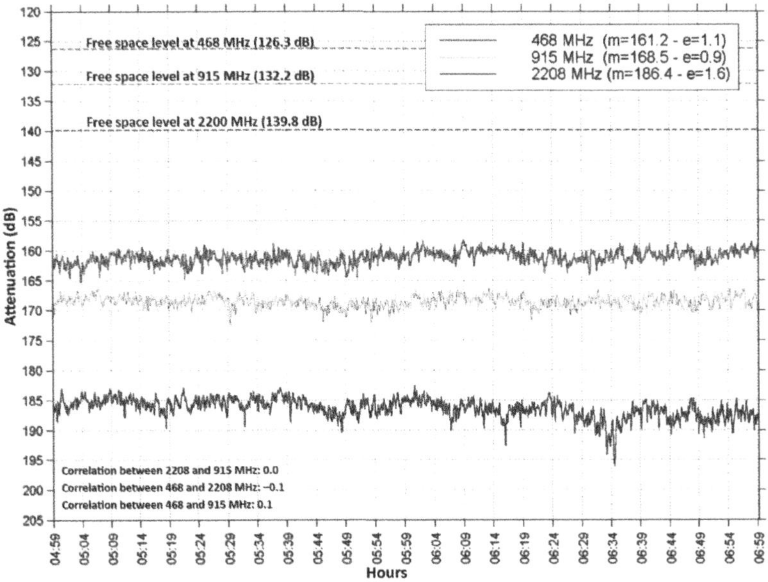

Figure 2.25. *Example of variation in the radioelectric field due to tropospheric scattering*

Rainscatter

Contrary to tropospheric scattering due to the permanent presence of heterogeneities of the refractive index within the atmosphere, scattering by rain only appears when it rains in the common volume of antenna beams.

If the diameter of the raindrops is relatively small compared to the wavelength, the scattering cross section of a raindrop (ratio of the power scattered to the surface power of the incident wave) is proportional to the square of the volume of the particle. It is given by Rayleigh's equation (Boithias 1983):

$$\sigma_i = \pi^5 \left(\frac{\varepsilon - 1}{\varepsilon + 2} \right)^2 \frac{D_i^6}{\lambda^4} \qquad [2.93]$$

where:

– ε is the permittivity of water;

– D_i is the diameter of the raindrop;

– λ is the wavelength.

The presence of the term λ^4 corresponds to a frequency selectivity (the blue color of the sky results from the scattering of white light by air molecules).

The total effective scattering cross section of all the drops contained in the unit volume V is:

$$\sigma_d = \sum_{i=1}^{N} \frac{N_i \sigma_i}{V} = \frac{\pi^5}{\lambda^4} \left(\frac{\varepsilon - 1}{\varepsilon + 2} \right)^2 \sum_{i=1}^{N} \frac{N_i D_i^6}{V} \qquad [2.94]$$

It is therefore proportional to:

$$\sum_{i=1}^{N} D_i^6 \qquad [2.95]$$

By defining the radar reflectivity factor Z by the relation:

$$Z = \frac{1}{V} \sum_{i=1}^{N} N_i D_i^6 \qquad [2.96]$$

as a result, the total scattering cross section becomes:

$$\sigma_d = \frac{\pi^5}{\lambda^4} \left(\frac{\varepsilon - 1}{\varepsilon + 2} \right)^2 Z \qquad [2.97]$$

The radar reflectivity factor Z is linked to the rain intensity R by the following relation:

$$Z = aR^b \qquad [2.98]$$

The coefficients *a* and *b* depend on the frequency and the polarization (ITU-R P.837). The reader will find in recommendation ITU-R P.837 models of rain intensity R_p exceeded for a given percentage of the average year, p, and at a given geographical location.

The ratio $\left(\frac{\varepsilon - 1}{\varepsilon + 2} \right)^2$, in the case of rain, is almost independent of the frequency, although the real and imaginary parts of ε vary greatly for frequencies above

104 GHz. For low frequencies and around 100 GHz, we will take it to be equal to 0.93. For ice, on the other hand, this does not exceed 0.2 (Boithias 1983).

The power received by scattering on a rain volume V located respectively at distances d_t and d_r from the transmitting and receiving antennas, and outside the direct path, is:

$$P_r = G_t G_r \frac{\lambda^2}{(4\pi)^3} \frac{\sigma_d V}{d_t^2 d_r^2} \quad [2.99]$$

Reflection on the upper layers

Reflection on the upper layers of the atmosphere acts as a signal repeater. These layers are not necessarily present throughout the link and the signal received depends on their angle of incidence and their surface condition. A small variation in these layers, both in surface condition and in position (altitude, inclination, etc.), can lead to large variations in the level of the received signal (Blanchard 1999). Figure 2.26 gives an example of variations in the received signal, observed on a 105 km link on the Alsace plain at 468, 915 and 2205 MHz in the presence of reflection phenomenon on the layers of the atmosphere (Blanchard 1999).

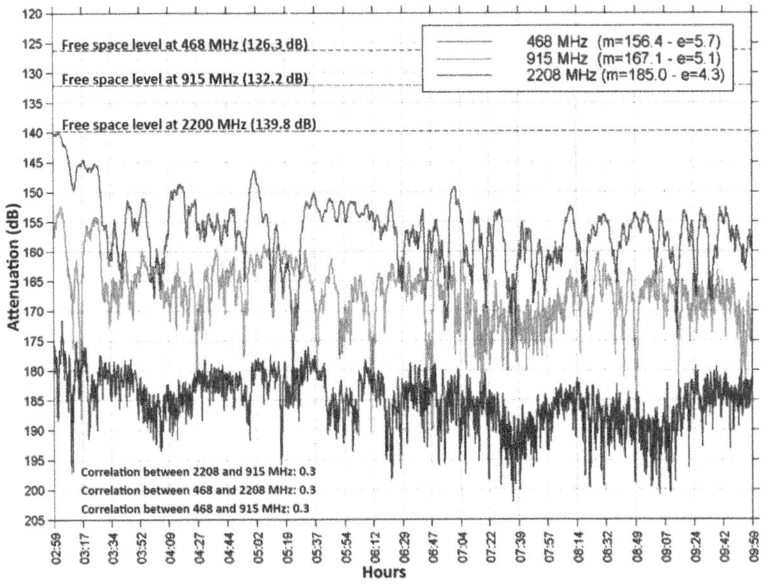

Figure 2.26. *Example of variation in the radioelectric field due to reflection on layers of the atmosphere*

Duct propagation

When the gradient of the refractive index is less than –157 N units/km, the paths of the rays are more curved than the terrestrial surface. The area of the atmosphere in which these super refractive conditions exist is called the ducting layer.

These layers have significant effects on the line-of-sight and cross-horizontal links. Indeed, in the presence of a duct, the notion of a radioelectric horizon no longer has any precise meaning because points located beyond the horizon can be reached. In addition, the level of the signal received during such a phenomenon can reach or even exceed the level in free space (Vilar *et al.* 1988; Shen 1995; Rana *et al.* 1993; Blanchard 1999). Ducting layers are therefore one of the main causes of interference between services using the same frequency range.

If the ducting layer is located low enough and the ground is sufficiently reflective, there will be a duct on the ground. On the other hand, if the layer is located high up, the paths of the rays are curved successively towards the Earth and towards space. The ray is therefore trapped between two altitudes and if it does not touch the ground, we speak of a high altitude or high duct.

The ducting layers mean that the rays from the transmitting antenna can cross at certain points in space. Thus, interference zones appear where the rays from the multiple paths intersect and zones where hardly any rays pass (the signal level is low) are known as *radioelectric holes.* The boundary between these two areas constitutes a causticity along which the level is very high. As the conditions of refraction vary over time, a point in space can be found alternately in one or the other of these regions. This results in sudden fluctuations in the received signal.

The thickness of a duct rarely exceeds a few hundred meters, but on the other hand a duct can extend over several hundred kilometers, especially above coastal areas or very humid areas. On the surface of the sea there can be an evaporation duct whose thickness is around a dozen meters, but it is present for a large percentage of time.

Figure 2.27 gives an example of variations in the received signal observed on a 105-kilometer link on the Alsace plain at 468, 915 and 2205 MHz in the presence of atmospheric duct (Blanchard 1999).

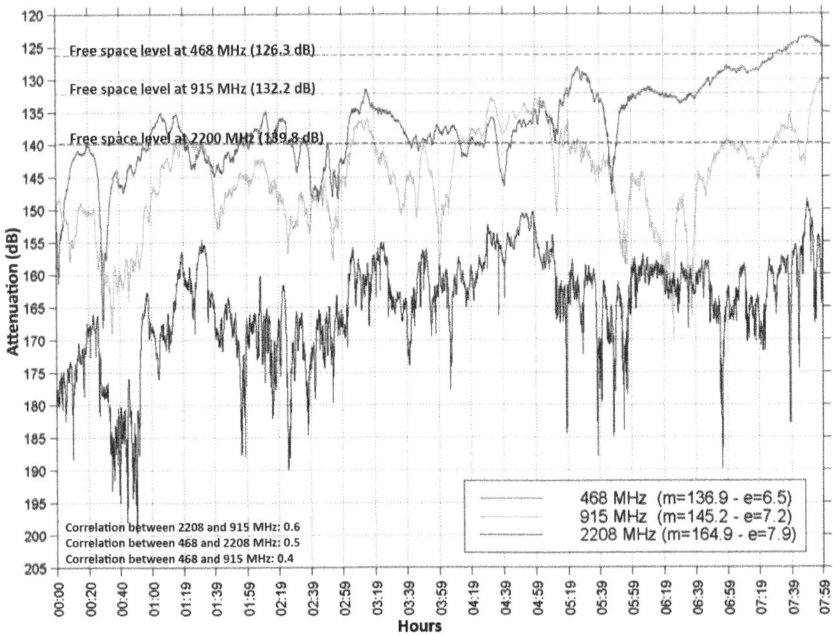

Figure 2.27. *Example of variation in radioelectric field due to the presence of ducts*

2.5.2. *Propagation in rural, suburban and urban areas*

The demand for low, medium and high-speed communications via the air has led manufacturers and operators to develop and deploy new technologies, for example GSM (Global System for Mobile Communications), UMTS (Universal Mobile Telecommunication Systems) and WIMAX (Worldwide Interoperability for Microwaves Access).

GSM, after RADIOCOM 2000, is the second-generation standard for mobile networks. It allows maximum bit rates of 9.6 kbit/s. It operates at 900 and 1,800 MHz. The access technique is based on a combination of TDMA time division and FDMA frequency division. It thus allows low volume voice and digital data transmissions (SMS, MMS).

UMTS is the 3rd generation standard for mobile networks. It is based on the W-CDMA multiple access technique, a so-called spread spectrum technique. The frequencies allocated for UMTS are 1,885–2,025 MHz and 2,110–2,200 MHz. The bit rate is different depending on the place of use: 144 kbits/s in rural areas (up to 500 km/h), 384 kbits/s in urban areas (up to 120 km/h) and 2 Mbits/s in a building. It allows multimedia content such as images, sounds and video.

WIMAX is a new radio access technology, based on a family of IEEE 802.16 standards, defining broadband links via the air and ensuring a high level of interoperability between different devices. Spectral efficiency is around 3 bits/s/Hz with the most efficient modulation. More efficient than WiFi, the objective of WIMAX is to provide high bit rate links (voice, data, video on demand) (over several tens of Mbits/s) over wireless networks even several tens of kilometers in radius. It thus allows wireless links between a base station (Base Transceiver Station, denoted BTS) and thousands of subscribers (Subscriber Station, denoted SS) through broadband "air" interfaces.

Then emerged 4th generation and 5th generation technology (4G and 5G) and associated objects. In addition to voice and internet access, it responds to use cases. There are many applications: health, smart city, agriculture, factory of the future, connected car.

Such systems operate at distances and at frequencies less than those used in microwave radio links. Variations in the refractive index of the atmosphere are negligible. Only the surrounding environment (buildings, vegetation, etc.) is considered. The dimensioning of such networks requires the development of specific propagation models in different environments (rural, suburban and urban and inside buildings, etc.).

2.5.2.1. *Geographic databases*

Predicting the propagation channel characteristics is based on knowledge of the environment described in geographic databases. They contain information concerning topography (relief), land use (wood, road, buildings, etc.), morphology, street axes, etc.

These different geographic databases result from processes using satellite (SPOT, HELIOS, etc.) or aerial photographs coupled with complex digitization processes.

Figures 2.28 and 2.29 give examples of the topography in the Perpignan region on the one hand and of the outlines of buildings in the city of Belfort on the other hand.

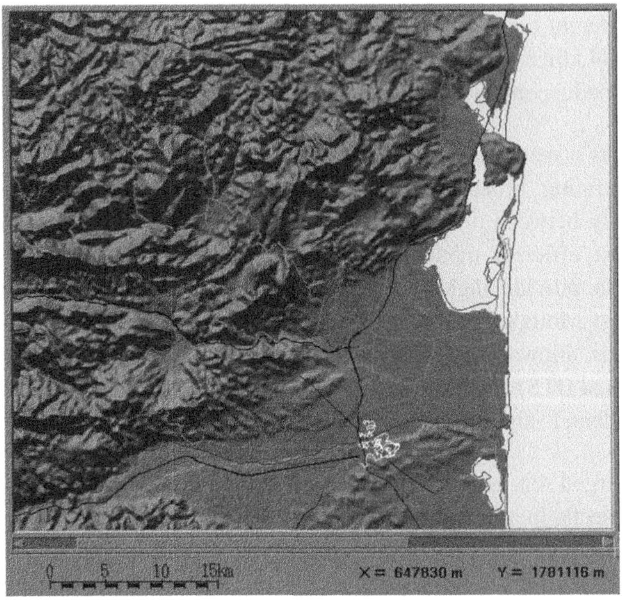

Figure 2.28. *Map of the topography (relief) in the Perpignan region, France*

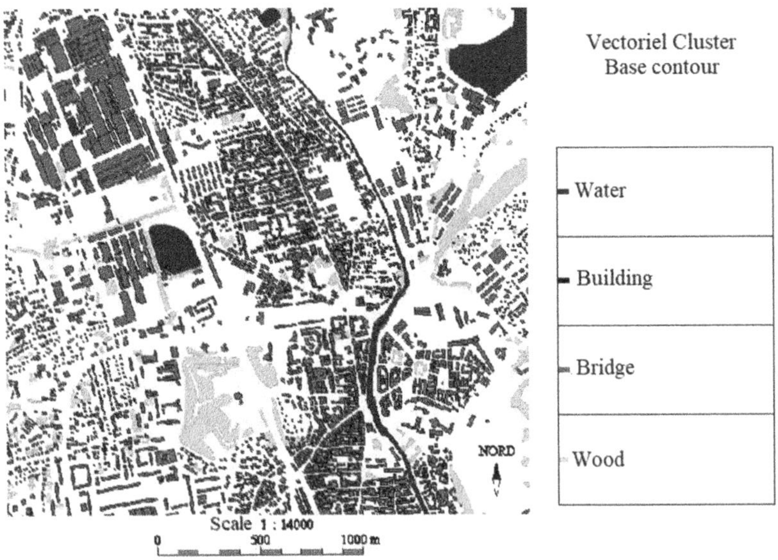

Figure 2.29. *Map of the topography (relief) in the Belfort region, France*

2.5.2.2. *Propagation models*

2.5.2.2.1. The rural model

The most widely used attenuation model in the macrocellular environment is the OKUMURA-HATA model (Okumura *et al.* 1968; Hata 1980).

Attenuation is given by the following equation:

$$\begin{aligned} A_p &= 69.55 + 26.16\log_{10} f - 13.82\log_{10} h_b + \ldots \\ &\ldots + \left(44.9 - 6.55\log_{10} h_b\right)\log_{10} d - a\left(h_m\right) \end{aligned} \qquad [2.100]$$

with:

– for small and medium-sized cities:

$$a\left(h_m\right) = \left(1.1\log_{10} f - 0.7\right)h_m - \left(1.56\log_{10} f - 0.8\right) \qquad [2.101]$$

– for big cities and $f \leq 200$ MHz:

$$a\left(h_m\right) = 8.29\left(\log_{10}\left(1.54h_m\right)\right)^2 - 1.1 \qquad [2.102]$$

– for big cities and $f \geq 400$ MHz:

$$a\left(h_m\right) = 3.2\left(\log_{10}\left(11.75h_m\right)\right)^2 - 4.97 \qquad [2.103]$$

where:

– h_b is the height of the base station;

– h_m is the mobile height.

This formulation of the attenuation is the result of measurements made near Tokyo. On the other hand, deviations have been noted when it is applied in cities whose characteristics are very different from this particular city (COST 231 1999).

The OKUMURA-HATA model for small and medium-sized cities has been extended to the 1,500–2,000 MHz frequency band as part of the COST 231 project: the model COST231-HATA:

$$\begin{aligned} A_p &= 46.3 + 33.9\log_{10} f - 13.82\log_{10} h_b + \ldots \\ &\ldots + \left(44.9 - 6.55\log_{10} h_b\right)\log_{10} d - a\left(h_m\right) + C_m \end{aligned} \qquad [2.104]$$

where:

– $a(h_m)$ is identical to the expression defined above;

– $C_m = 0$ dB in small and medium cities as well as in urban areas;

– $C_m = 3$dB in large cities.

Its validity field is described by the following parameters:

– frequency range: 500–1,500 MHz, with possible extension to 2,000 MHz (COST 231 1999);

– height of base stations varying between 30 and 200 m in height;

– height of reception antennas between 1 and 10 m in height;

– distance varying between 1 and 20 km.

However, this model is not suitable for lower base station antennas and mountainous or wooded environments. In order to overcome these various drawbacks, the following model has been proposed (Erceg 1999). It applies to the following environments:

– type A terrain: mountainous terrain covered with medium and high-density forests. It is characterized by a strong attenuation;

– type B terrain: intermediate environment between type A and type C;

– type C terrain: flat with low vegetation cover.

For a very small distance d_0, the median of the propagation loss is given by the following equation:

$$L = A + 10\gamma \log_{10}\left(\frac{d}{d_0}\right) + s \qquad d > d_0 \qquad [2.105]$$

where:

– $A = 20\log_{10}\left(\frac{4\pi d}{\lambda}\right)$;

– λ is the wavelength;

– γ is an attenuation coefficient;

– $d_0 = 100$;

– s is a variable representing shadowing effects.

Term A can also be written in logarithmic form:

$$A = 32.4 + 20\log_{10} f + 20\log_{10} d \quad [2.106]$$

where:

– f is the frequency in MHz;

– d is the distance in km.

The coefficient γ is given by the equation:

$$\gamma = \left(a - b.h_b + \frac{c}{h_b} \right) \quad 10m < h_b < 100m \quad [2.107]$$

where h_b is the height of the base station.

The values of coefficients a, b and c are given in Table 2.2.

Coefficient	Type A	Type B	Type C
a	4.6	4	3.6
b	0.0075	0.0065	0.005
c	12.6	17.1	20

Table 2.2. *Values of coefficients A, B and C*

The variable *s*, characterizing the shadowing effects, follows a normal log distribution. Typical values of standard deviation range from 8.2 to 10.6 dB, depending on the density of the vegetation (Erceg 1999).

This model is applied more particularly for frequencies around two gigahertz and reception antenna heights around two meters. Corrective terms are introduced in the attenuation equation in order to extend its validity field to other frequencies, and to receiving antenna heights between two and ten meters. The median of the propagation loss is given by the following equation:

$$L_{\text{modified}} = L + \Delta L_f + \Delta L_h \quad [2.108]$$

where:

– L is the attenuation given in the previous expression;

– ΔL_f is the corrective term due to the frequency;

– ΔL_h is the corrective term due to the height of the reception antenna.

The frequency correction term is given by the following equation (Jakes and Reudink 1967; Chu and Greenstein 1999):

$$\Delta L_f = 6\log_{10}\left[\frac{f}{2000}\right] \quad [2.109]$$

where f is the frequency in MHz.

The corrective term due to the height of the receiving antenna is given by the following equation:

$$\Delta L_h = 10.8\log_{10}\left(\frac{h}{2}\right) \text{ for type A and B terrain} \quad [2.110]$$

$$\Delta L_h = -20\log_{10}\left(\frac{h}{2}\right) \text{ for type C terrain} \quad [2.111]$$

where h is the antenna height between 2 and 10 meters.

2.5.2.2.2. Suburban and urban models

In addition to the OKUMURA-HATA model written above, two models were chosen more specifically to characterize the channel propagation in urban and suburban areas.

The first, the ECC-33 (Electronic Communication Committee) model, was developed by the CEPT (The European Conference of Postal and Telecommunications Administrations). It applies more particularly in the range of frequencies 3.4 to 3.8 GHz (Electronic Communication Committee 2003). Not using geographic databases, it finds a more particular application in the dimensioning of a network.

The second, the Walfish-Ikegami model, was developed as part of the COST 231 action (COST 231 1999). It is a semi-empirical model adapted to the context of small urban and suburban cells (coverage radius of a few kilometers, antennas above roof level). It is more suitable for GSM (900 MHz) and DCS (1,800 MHz) engineering with an emission between 5 and 15 meters above roof level. The propagation mainly takes place above roofs to finally plunge into the street where the mobile moves (Figure 2.30). It takes into account the free space loss, the loss related to multiple edge diffraction and the loss related to the last diffraction.

The ECC-33 model

The propagation loss is described by the following equation:

$$PL = A_{fs} + A_{bm} - G_b - Gr \qquad [2.112]$$

where:

- A_{fs} is the free space loss;
- A_{bm} is the median propagation loss;
- G_b is the height gain factor of the base station antenna;
- G_r is the height gain factor of the CPE terminal antenna.

These different parameters are defined as follows:

$$A_{fs} = 92.4 + 20\log_{10}(d) + 20\log_{10}(f) \qquad [2.113]$$

$$G_b = \log_{10}\left(\frac{h_b}{200}\right)\left\{13.958 + 5.8\left[\log_{10} d\right]^2\right\} \qquad [2.114]$$

$$G_r = \left[42.57 + 13.7\log_{10}(f)\right]\left[\log_{10}(h_r) - 0.585\right] \qquad [2.115]$$

where:

- f is the frequency of the electromagnetic wave (GHz);
- d is the distance between the base station and the terminal station (km);
- h_b is the height of the base station (m);
- h_r is the height of the terminal station (m).

Walfish-Ikegami model

Propagation attenuation is given by the following equation:

$$L_b = L_0 + L_{rts} + L_{msd} \qquad [2.116]$$

where:

– L_0 is the free space loss;

– L_{rts} is the attenuation due to diffraction on the last edge;

– L_{msd} is the attenuation due to multiple edge diffraction.

Figure 2.30 gives a schematic representation of the "transmitter-receiver" profile and a definition of the different parameters used in the model.

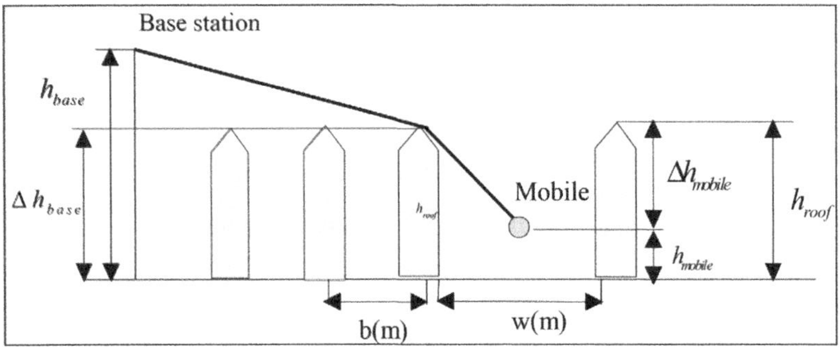

Figure 2.30. *Schematic representation of the "transmitter-receiver" profile*

Free space loss L_0 is given by the fundamental equation of telecommunications:

$$L_0 = 32.4 + 20\log_{10} d_{km} + 20\log_{10} f_{MHz} \quad [2.117]$$

The loss associated with the last diffraction is: when $h_{mobile} < h_{roof}$:

$$L_{rts} = -16.9 - 10\log_{10}(w_m) + 10\log_{10}(f_{MHz}) + \ldots$$
$$\ldots + 20\log_{10}(\Delta h_m) + L_{ori} \quad [2.118]$$

$$L_{rts} = 0 \qquad \text{if } L_{rts} < 0$$

where:

– $L_{ori} = -10 + 0.354\varphi_{degree}$; $0 \leq \varphi_{degree} \leq 35$ [2.119]

– $L_{ori} = 2.5 + 0.075(\varphi_{degree} - 35)$; $35 \leq \varphi_{degree} \leq 55$ [2.120]

– $L_{ori} = 4.0 + 0.114\left(\varphi_{degree} - 55\right)$; $55 \le \varphi_{degree} \le 90$ [2.121]

– $\Delta h_{mobile} = h_{roof} - h_{mobile}$ [2.122]

– w_m is the street width (m);

– f_{Mhz} is the wave frequency (MHz);

– φ_{degree} is the angle (in °) between the street axis and the direction of the incident wave (Figure 2.31).

Loss due to multiple diffraction on the roofs:

$$L_{msd} = L_{beh} + k_a + k_d \log_{10} d_{km} + k_f \log_{10} f_{MHz} - 9\log_{10} b_m \quad [2.123]$$

$$L_{msd} = 0 \qquad \text{if } L_{msd} < 0$$

where:

– $L_{beh} = 18\log_{10}\left(1 + \Delta h_{base}\right)$ if $h_{base} > h_{roof}$;

– $L_{beh} = 0$ if $h_{base} \le h_{roof}$;

– $k_a = 54$ if $h_{base} > h_{roof}$;

– $k_a = 54 - 0.8\Delta h_{base}$ if $d_{km} \ge 0.5\, and\ h_{base} \le h_{roof}$;

– $k_a = 54 - 0.8\Delta h_{base} \dfrac{d_{km}}{0.5}$ if $d_{km} < 0.5\, and\ h_{base} \le h_{roof}$;

– $k_d = 18$ if $h_{base} > h_{roof}$;

– $k_d = 18 - 15\dfrac{\Delta h_{base}}{h_{roof}}$ if $h_{base} \le h_{roof}$;

– $k_f = -4 + 0.7\left(\dfrac{f_{MHz}}{925} - 1\right)$ for an average city and a suburban area (vegetation with average density);

– $k_f = -4 + 1.5\left(\dfrac{f_{MHz}}{925} - 1\right)$ for dense urban areas;

– b_m(m) is the separation between buildings.

Note that $\Delta h_{base} = h_{base} - h_{roof}$ (the different parameters here are in meters).

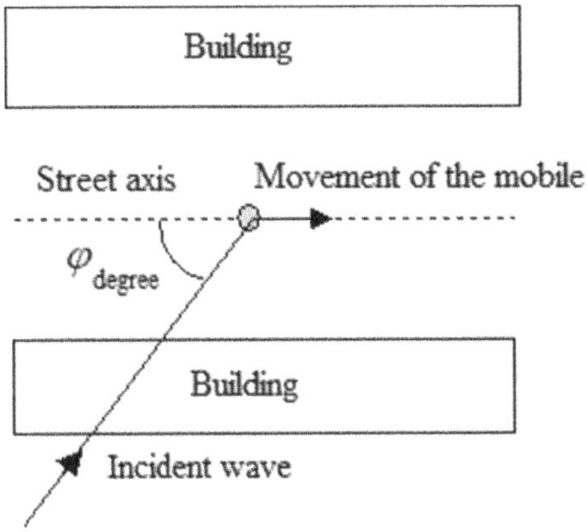

Figure 2.31. *Definition of the angle between the street axis and the direction of the incidence angle*

Its field of validity is described by the following parameters:

– frequency range: 800–2,000 MHz;

– base station height varying between 4 and 50 meters;

– reception antenna height between 1 and 3 meters;

– distance varying between 20 and 5 kilometers.

This is consistent with experimental measurements in urban and suburban environments characterized by a uniform height of buildings (the standard deviation between measurements and the projection is around 6 dB). In addition, it agrees with the previous model in the case of type C terrain (plain with low vegetation cover). It thus allows continuity between the two models. A corrective term considers the height of the receiving antenna. This model can be used in both urban and suburban environments.

Radioelectric engineering

Carried out using specific tools, radioelectric engineering involves determining the characteristics of channel propagation considering the different parameters in the presence of the actual environment characterized by a geographic database (topography, morphology, contours of buildings, street axis, etc).

These are computer-assisted cellular network planning tools. They provide the user with the means to design, study and optimize radioelectric engineering. In particular, they predict the field received from various transmitters using a geographic database and validated propagation models on a set of measurements in the field. They also make it possible to analyze the interference caused by the reuse of frequency.

Figure 2.32 gives an example of a "transceiver" profile.

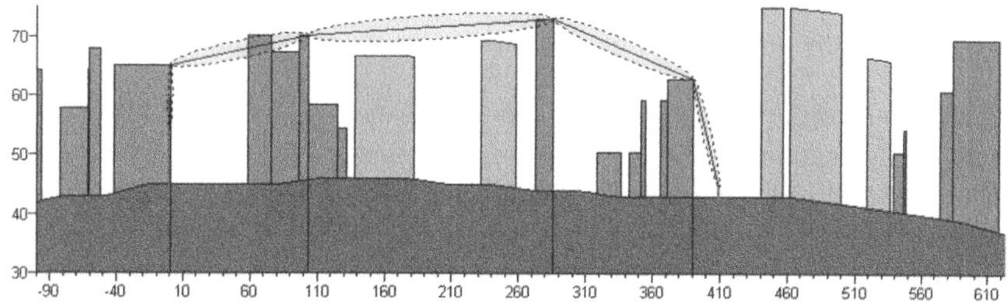

Figure 2.32. *Example of a "transceiver" profile*

Figure 2.33 gives an example of radioelectric coverage.

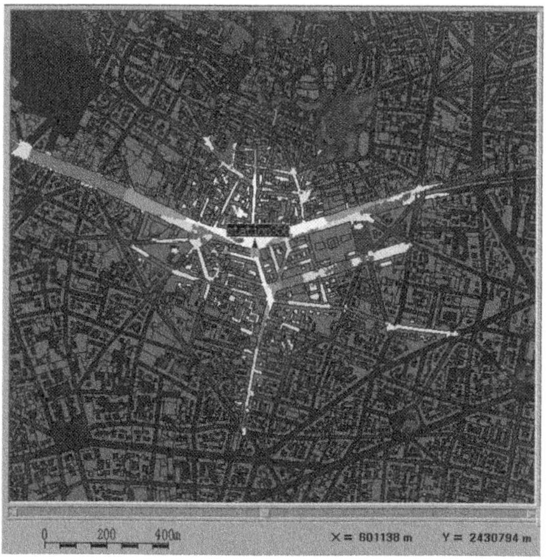

Figure 2.33. *Example of urban coverage*

2.5.3. *Propagation within buildings*

A distinction is made between the penetration loss in buildings on the one hand and the propagation loss inside buildings on the other hand.

2.5.3.1. *Penetration models*

Penetration attenuation in a building is defined as the loss of power experienced by the electromagnetic field between the outside around the building and one or more positions inside the building. It is calculated by comparing the external field and the field in the rooms of the building where the receiving mobile is located.

The parameters affecting the penetration attenuation values are multiple and their effects mostly intermingle. Among these many parameters, the following conventional ones are generally distinguished: the surrounding environment; the reception depth in buildings; the angle of incidence; the reception height, more commonly called the "floor effect"; the transmitter-receiver distance; height of transmitting antenna; frequency; type of material traversed; etc. (Sizun 2003).

The most classic models are inspired by the Motley Keenan model (Motley and Keenan 1998) used in propagation inside buildings. They predict penetration attenuation based on parameters such as:

– the distance between the transmitter and the outside wall of the building where the receiver is located;

– the distance between the exterior wall and the receiver;

– the number of internal walls along the emitter-receiver profile;

– the floor effect;

– the loss of energy off the exterior wall of the building;

– the loss of energy off the internal walls.

The propagation loss (L) is expressed as a sum of the losses in free space (L_0), losses due to obstacles crossing the direct ray (slabs, walls, doors, windows), of a constant (L_c) (Motley and Keenan 1998). The database can differentiate between the various obstacles associated with a particular attenuation value. This is the most used model:

$$L = L_0 + L_c + \sum_{j=1}^{N} N_j L_j + N_f L_f \qquad [2.124]$$

where:

– N_j is the number of type j walls crossed;

– L_j are the losses due to type j walls;

– N is the number of different wall types;

– N_f is the number of slabs crossed;

– L_f are the losses per slab.

Typical values of losses as a function of materials (exterior walls) in the 1–2 GHz frequency band are summarized in Table 2.3.

Materials	Loss (dB)
Porous concrete	6.5
Reinforced glass	8.0
Concrete (30 cm)	9.5
Wall made of thick concrete (25 cm) with large windows	11.0
Wall made of thick concrete (25 cm) without glass	1.0
Thick wall (> 20 cm)	15
Slab	23

Table 2.3. *Transmission losses of different construction materials (exterior walls) in the 1–2 GHz band*

2.5.3.2. *Models inside buildings*

2.5.3.2.1. Geographic databases

The characteristics of the propagation channel are predicted based on knowledge of the indoor environment generally described in the form of facets. Several types of materials can be considered: load-bearing walls, partitions, slabs, doors, windows, etc. Each of them is characterized by its dielectric parameters (permittivity, permeability) and its thickness.

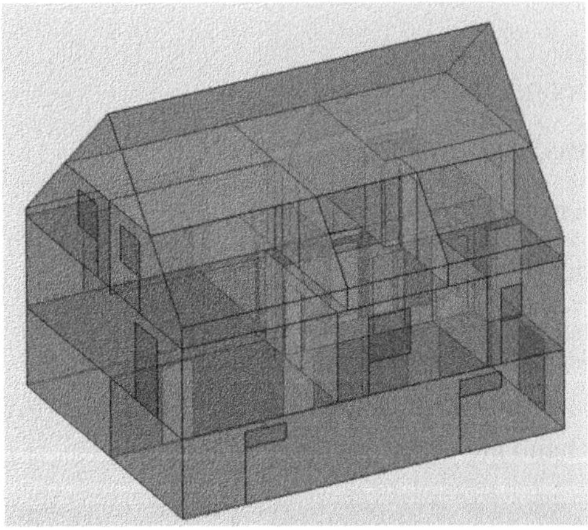

Figure 2.34. *Example of a representation of a residential environment*

The propagation of radioelectric waves inside buildings essentially depends on the nature of the environment (closed offices, open offices, station or airport halls, corridors, residences, etc.). It is characterized by the presence of numerous routes (multi-paths); the predominant propagation mechanisms are reflection, transmission, diffraction and diffusion. Many models, both statistical and deterministic, for characterizing propagation loss exist in the literature. We detail below only the statistical models which do not require detailed building databases. Particular emphasis is placed on the following models: Motley-Keenan, ITU-R, COST 231, COST 259, MWF, IEEE P802.11.

2.5.3.2.2. Motley-Keenan model

The propagation loss (L), like the penetration loss inside buildings, is expressed by a sum of the losses in free space (L_0), losses due to obstacles crossing the direct ray (slabs, walls, doors, windows), of a constant (L_c) (Motley and Keenan 1998). The database can differentiate between the various obstacles associated with a particular attenuation value. This is the most used model:

$$L = L_0 + L_c + \sum_{j=1}^{N} N_j L_j + N_f L_f \qquad [2.125]$$

where:

– N_j is the number of type j walls crossed;

– L_j losses due to type j walls;

– N is the number of different wall types;

– N_f is the number of slabs crossed;

– L_f are the losses per slab.

Typical values for losses by materials (exterior walls) in the 1–2 GHz frequency band are summarized in Table 2.4 (COST 231 1999).

Materials	Losses (dB)
Plasterboard	3
Wood	3
Window	2
Brick wall less than 14 cm thick	4
Wall made of concrete less than 10 cm thick	13.0
Thick wall (> 20 cm)	17
Slab	23
Metal wall	30

Table 2.4. *Transmission losses of different construction materials (interior walls in the 1–2 GHz band)*

2.5.3.2.3. ITU-R model

Described in Recommendation ITU-R P.1238, the total propagation loss is written as follows:

$$L_{total} = 20\log_{10} f + N\log_{10} d + L_f(n) - 28 \qquad [2.126]$$

where:

– N is the distance power loss coefficient;

– f is the frequency (MHz);

– d is the distance (m) separating the base station from the terminal receiver (d > 1 m);

– L_f is the floor penetration loss factor (dB);

– n is the number of stages between the base station and the receiver ($n \geq 1$).

Tables 2.5 and 2.6 give some examples of N and floor penetration loss factor in an office and commercial environment for different values of the frequency.

Frequency	Offices	Commercial buildings
900 MHz	33	20
1.2–1.3 GHz	32	22
1.8–2 GHz	30	22
4 GHz	28	22
5.2 GHz	31	–

Table 2.5. *Distance power loss coefficient*

Frequency	Offices	Commercial buildings
900 MHz	9 (1 floor) 19 (2 floors) 24 (3 floors)	9 (1 floor) 19 (2 floors) 24 (3 floors)
1.8–2 GHz	15 + 4 (n – 1)	6 + 3 (n – 1)
5.2 GHz	16 (1 floor)	16 (1 floor)

Table 2.6. *Floor penetration loss factor ($n \geq 1$)*

2.5.3.2.4. COST 231 models

The European project COST 231 (EURO-COST-231 1999) developed three empirical models:

– linear attenuation model;

– one-slope model;

– multi-wall model.

Linear attenuation model

The additional free space loss depends linearly on the transceiver distance:

$$L = L_{FS} + \alpha d \qquad [2.127]$$

where:

– α is the attenuation coefficient (dB per meter);

– L_{FS} is the loss of free space between the transmitter and the receiver (dB).

One-slope model

The one-slope model considers a linear dependence between the attenuation and the logarithm of the transmitter-receiver distance:

$$L = L_0 + 10n\log_{10}(d) \text{ in dB} \qquad [2.128]$$

where:

– L_0 is the attenuation when d=1 m;

– n is the power loss coefficient;

– d is the transceiver distance (m).

The value of the power attenuation coefficient n depends on the environment.

Multi-wall model

The multi-wall model considers both the free space attenuation and the attenuation linked to the crossing of walls and floors by the direct path. As the attenuation due to the crossing of floors is not linear to the number of floors, an empirical factor b has been introduced into the expression of the attenuation:

$$L = L_{FS} + L_c + \sum_{j=1}^{I} k_{wi} L_{wi} + k_f^{\left[\frac{k_f+2}{k_f+1} - b\right]} L_f \qquad [2.129]$$

where:

– L_{FS} is the loss of free space between the transmitter and the receiver (dB);

– L_c is the attenuation constant (dB);

– k_{wi} is the number of type i walls crossed;

– k_f is the number of floors crossed;

– L_{wi} is the attenuation due to the crossing of type i walls (dB);

– L_f is the attenuation associated with crossing a floor (dB);

– b is an empirical parameter;

– I is the number of wall types.

Two types of walls are available:

– light walls (L_{w1}), which are not load-bearing walls (plasterboard, chipboard, etc.) or thin walls whose thickness is less than ten centimeters (light concrete wall);

– heavy walls (L_{w2}), such as load-bearing walls or walls whose thickness is more than ten centimeters (concrete, brick, etc.).

Tables 2.7a and 2.7b give the values of the different parameters of the models described above. They were obtained from 1,800 MHz measurements in four types of buildings (dense, open, wide and corridor).

Building	Linear model	One-slope model	
	α	L_0 (dB)	n
Dense	–	33.3	4
1 floor	0.62	33.3	4
2 floors	–	21.9	5.2
Multi-floor	2.8	44.9	5.4
Open	0.22	42.7	1.9
Wide	–	37.5	2
Corridor	–	39.2	1.4

a)

Building	Multi-wall model			
	L_{w1} (dB)	L_{w2} (dB)	L_f (dB)	b
Dense	–	–	–	–
1 floor	3.4	6.9	18.3	0.46
2 floors	–	–	–	–
Multi-floor	–	–	–	–
Open	3.4	6.9	18.3	0.46
Wide	3.4	6.9	18.3	0.46
Corridor	3.4	6.9	18.3	0.46

b)

Table 2.7. *a) Coefficients of the linear model and one slope model obtained from measurements at 1,800 MHz; b) coefficients of the multi-wall model obtained from measurements at 1,800 MHz*

2.5.3.2.5. COST 259 model

The COST 259 model (COST 259 2000) is an improved version of the multi-wall COST 231 model:

$$L = L_{FS} + \sum_{i=1}^{I} k_{wi}^{\left[\frac{k_{wi}+1.5}{kwi+1}-b_{wi}\right]} L_{wi} \qquad [2.130]$$

This model does not consider the path loss between floors. It is only valid for attenuation calculations on the same floor.

The term b_{wi} depends on L_{wi} per the following equation:

$$b_{wi} = -0.064 + 0.0705 L_{wi} - 0.0018 L_{wi}^2$$

Table 2.8 gives the parameters of the multi-wall model at 5 GHz.

Materials	L_{w1} (dB)	L_{w2} (dB)
Plasterboard	3.4	0.15
Mixed walls (plaster, concrete, glass)	8.4	0.4
Concrete	11.8	0.52

Table 2.8. *Penetration loss and non-linear parameter at 5 GHz*

2.5.3.2.6. Multi-Wall and Floor (MWF) model

This model, proposed by Lott *et al.* (Lott and Forkel 2001), considers the linear non-dependence between the total attenuation and the number of walls or floors crossed in the same category. It takes into account the number of floors crossed. It is written in the following form:

$$L_{MWF} = L_0 + 10n\log_{10}(d) + \sum_{i=1}^{I}\sum_{k=1}^{K_{wi}} L_{wik} + \sum_{j=1}^{J}\sum_{k=1}^{Kfj} L_{fjk} \qquad [2.131]$$

where:

– L_0 is the loss at a distance of 1 m (dB);

– n is the power loss coefficient;

– d is the transceiver distance (m);

– L_{wik} is the attenuation due to the crossing of kth type i wall (dB);

– L_{fjk} is the attenuation due to the crossing of kth type j floor (dB);

– l is the number of wall types;

– J is the number of floor types;

– K_{wi} is the number of type i walls crossed (dB);

– K_{fj} is the number of type j floors crossed (dB).

The attenuation coefficient n has values between n = 1.96 and n = 2.03.

Table 2.9 gives the values of the MWF model parameters in the case of concrete walls.

Type of material	Thickness	K = 1	K = 2
Concrete	10 cm	Lw11 = 16	Lw12 = 14
	20 cm	Lw21 = 29	Lw22 = 24
Dry concrete	24 cm	Lw21 = 35	Lw22 = 29
Porous concrete	24 cm	Lw31 = 4	Lw32 = 26

Table 2.9. *Parameters of the MWF model in the case of concrete at 5.2 GHz*

2.5.3.2.7. The IEEE P802.11 model

The IEEE 802.11 Task Group n has developed different models, more particularly suited to domestic and small office environments, defined according to the spread of impulse response delays. (A-E model) (Erceg 2004).

The propagation loss depends on the transmitter-receiver distance relative to a distance d_{BP} (breakpoint distance):

$$L(d) = L_{FS}(d) \qquad d \le d_{BP}$$

$$L(d) = L_{FS}(d_{BP}) + 35\log_{10}\left(\frac{d}{d_{BP}}\right) \qquad d > d_{BP} \qquad [2.132]$$

where:

– d is the transmitter-receiver distance (m);

– L_{FS} is the free space loss.

The value of the distance d_{BP} varies from one model to another. Tables 2.10 and 2.11 give, for each model, each environment, the spread of the delays and the value of the distance d_{BP} (Table 2.10) as well as the standard deviations of shadowing effects in LOS ($d < d_{BP}$) and NLOS ($d > d_{BP}$) (Table 2.11).

Model	Environment	Delays (ns)	d_{BP} (m)
A	1 path	0	5
B	Residential	15	5
C	Small offices	30	5
D	Typical offices	50	10
E	Large offices	100	20
F	Wide spaces	150	30

Table 2.10. *Values for delay and distance d_{BP} according to the different types of environment*

Model	Environment	σ: LOS	σ: NLOS
A	1 path	3	4
B	Residential	3	4
C	Small offices	3	5
D	Typical offices	3	5
E	Large offices	3	6
F	Wide spaces	3	6

Table 2.11. *Values of the standard deviations of the shadowing according to the different types of environment (σ is the standard deviation)*

Comparing the different models at 2.4 and 5 GHz shows that the Motley-Keenan, Multiwall (COST 231 and 259) and IUT-R models give similar results. The COST 231 one-slope model is more pessimistic. However, it appears that the model proposed by the ITU-R is the most robust. It is based on an *nlogd+c* type equation, where n and c are parameters resulting from numerous measurements carried out in many environments.

2.5.3.2.8. Radioelectric engineering

Based on a 3D description of the environment, this is based on methods of ray launching and ray tracing.

The ray launching technique consists of launching a set of rays from the transmitter in all directions and searching for the set of rays that reach the receiver, after having considered all the physical phenomena that occur during their propagation: reflection, diffraction, penetration or crossing of vegetation, etc. The ray tracing technique is based on the determination of the different rays that travel through the space between a transmitter and a receiver, using the optical image method. Fresnel relations and the uniform theory of diffraction (UTD) are, respectively, implemented during the different reflections and reflections.

These techniques make it possible to predict the field, to determine the different rays which traverse the space between a transmitter and a receiver, to calculate the impulse response and the directions of the incoming and outgoing rays of the propagation channel.

Figure 2.35 gives an example of the coverage of the radioelectric field in a residential environment.

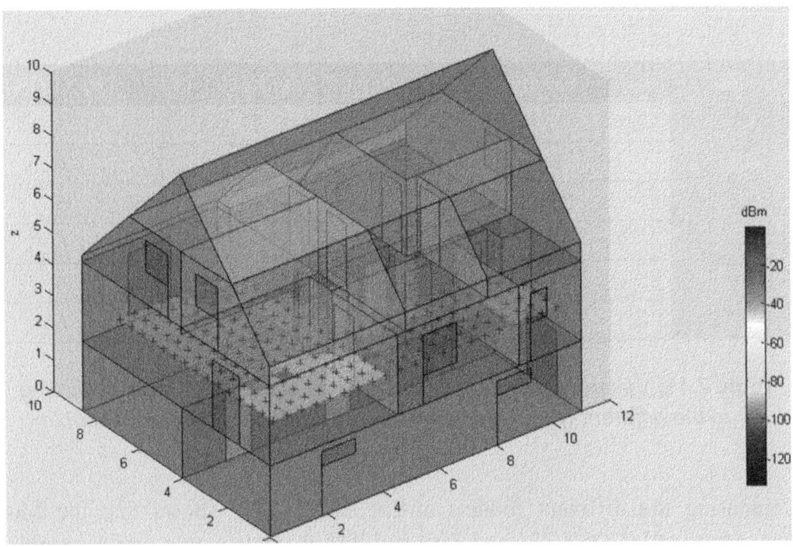

Figure 2.35. *Example of radioelectric coverage in a residential environment*

2.5.3.2.9. Furniture effects

Based on measurements made at 5.8 GHz, Cuinas and Sanchez (Cuinas and Sanchez 2004) have demonstrated the impact of furniture on attenuation. The basic equation used is an exponentially decreasing function:

$$k(dB) = a(dB) + 20n\log_{10}(d) \quad [2.133]$$

where:

– d is the distance between the transmitter and the receiver (m);

– n is the power loss coefficient;

– a is a constant;

– k(dB) is the attenuation.

The values of parameters a and n deduced from the different experiments are summarized in Table 2.12. Attenuation increases more quickly with distance in a furnished room than in an empty room.

Environment	a(dB)	n
Empty room	+ 46.02	+ 0.14
Furnished room	+ 37.72	+ 0.80
Square furnished room (same surface)	+ 37.72	+ 0.58
Large furnished room	+ 37.72	+ 1.05

Table 2.12. *Values of parameters a and n in different environment configurations*

The validity range of these equations extends more particularly to a distance of 4.5 to 7 meters, the range of distances considered during the experiment carried out.

2.5.3.2.10. Effects of people

From the distributions of measured signal levels at 2.4 GHz, empirical relationships have been deduced (Klepal *et al.* 2004). They describe the average attenuation, the standard deviations and the probability of having a line-of-sight link between the transmitter and the receiver.

The average additional loss compared to free space due to the movement of people (dB), the standard deviation and the probability of having a line-of-sight link are given respectively by the following equations:

$$\mu_s\left(l,\rho_p\right)=\left(3l\rho_p\right)^{0.7} \quad [2.134]$$

$$\sigma_s\left(l,\rho_p\right)=\log_7\left(55l\rho_p+1\right)+0.5 \quad [2.135]$$

$$A\left(l,\rho_p\right)=\left(1-\rho_p\right)^{0.2l} \quad [2.136]$$

where:

– l is the dimension of the area within which people move (m);

– ρ_p is the density of people (number of people per occupied area) (m^{-2}).

2.5.4. *Broadband propagation*

Due to the presence of multiple paths, the radioelectric channel propagation is frequency-selective and temporally and spatially variable due to variations in the radio environment. A broadband model represents its behavior in a given configuration (frequency, bandwidth, mobility (fixed, pedestrian, car, etc.)) and a given environment (indoor, outdoor, rural, urban, etc.). It takes into account temporal dispersion of the impulse response (multiple paths) caused by the many phenomena involved in the propagation process (reflection, transmission, diffraction, scattering, etc.). It is represented by a function representing the power delay profile (PDP) known as the impulse response (h(τ)). It is the response at time t to an impulse emitted at time t-τ. (Figure 2.8). It is typical to qualify the selectivity of the channel by parameters deduced from the average power profile of the impulse response. The most used are the average delay, the delay spread, the interval of the delays, the window of the delays and the band of coherence of the channel (Failly 1989).

There are two types of models, models for simulations and models for prediction:

– in the case of simulation, the channel is reproduced with its characteristics of temporal and spatial variability in a given environment from measurements, for example, to evaluate its effects before and after correction on the quality of the digital signal. The simulation models are intended to be integrated into digital transmission simulation chains with COSSAP type software (COS 1997);

– in the case of forecasting, a minimum number of characteristics is predicted from environmental data. Broadband predictive models are intended for engineering tools. The role of the predictive model is to predict the impulse response of the channel or, at least, of the parameters which directly influence the quality of transmission (number of paths, spread).

We distinguish between path models and geometric models.

2.5.4.1. *Path models*

These represent the impulse response by a limited number of discrete paths in different propagation environments (RA: rural, TU and BU: urban, HT: very hilly). Each path is defined by a power relative to that of the path of higher power (relative power in dB), a delay time and a type of Doppler spectrum (Rayleigh, Gauss, Rice). The discretization of the temporal space (delay time) causes a frequency periodicity of the model. It is, however, not very usable in the case of frequency shift.

Six- and twelve-path models were defined in four different environments:

– a rural environment (RA), for which a dominant path emerges. It has a poorly selective channel propagation;

– an urban environment of medium selectivity (UT) and high selectivity, when the link is respectively moderately and strongly obstructed (BU);

– a very hilly environment (HT).

Tables 2.13 and 2.14 give an example of an impulse response of six paths in an urban environment and twelve paths in a hilly environment. The DS parameter represents the standard deviation of the delay.

Path number	Delay (μs)	Relative power (dB)	Doppler	DS (μs)
1	0.0	– 3	Rayleigh	2.4
2	0.4	0	Rayleigh	2.4
3	1.0	– 3	Gauss 1	2.4
4	1.6	– 5	Gauss 1	2.4
5	5.0	– 2	Gauss 2	2.4
6	6.6	– 4	Gauss 2	2.4

Table 2.13. *Impulse response model in an urban environment (BU)*

Path number	Delay (μs)	Relative power (dB)	Doppler	DS (μs)
1	0.0	− 10	Rayleigh	5.0
2	0.2	− 8	Rayleigh	5.0
3	0.4	− 6	Rayleigh	5.0
4	0.6	− 4	Gauss 1	5.0
5	0.8	0	Gauss 1	5.0
6	2.0	0	Gauss 1	5.0
7	2.4	− 4	Gauss 2	5.0
8	15.0	− 8	Gauss 2	5.0
9	15.2	− 9	Gauss 2	5.0
10	15.8	− 10	Gauss 2	5.0
11	17.2	− 12	Gauss 2	5.0
12	20.0	− 14	Gauss 2	5.0

Table 2.14. *Impulse response model in hilly environment (HT)*

Figure 2.36 represents a GSM TU channel with 12 paths.

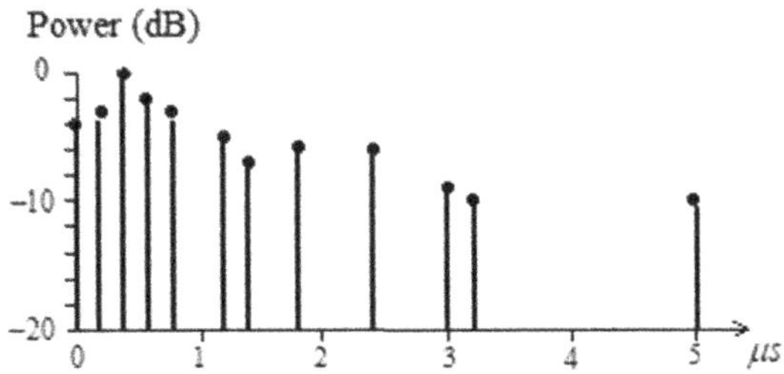

Figure 2.36. *Schematic representation of a GSM TU channel with 12 paths*

2.5.4.2. *Geometric models*

The concept of geometric modeling is based on the relationship that there can be between angular and temporal power profiles, and the location of reflectors and/or clusters in the propagation environment.

Indeed, as illustrated in Figure 2.37, the shape of the impulse response is directly related to the position of the dominant cluster, compared to the base station and the mobile. A path is made up of several rays, in order to consider the notions of spatial and temporal shift.

These rays are defined by a radio propagation path containing two successive bounces: one on a distant reflector or diffuser, the other on a diffuser close to the mobile (in order to obtain realistic Doppler spectra).

The direct path between the transmitter and the receiver, for its part, is composed solely of rays scattered by reflectors close to the mobile (Figure 2.37).

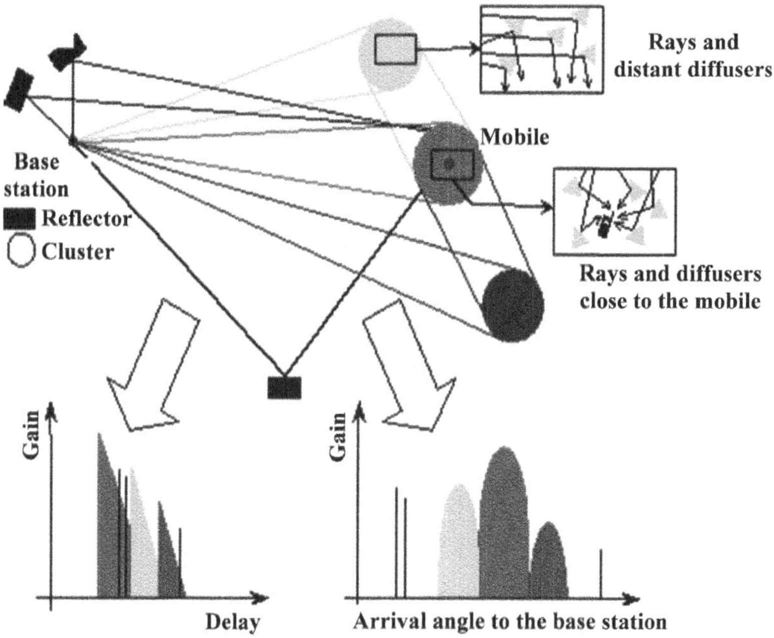

Figure 2.37. *Relations between the position of reflectors and diffusers in the propagation environment; shape of the power time profile*

The modeling phase involves identifying the position of the predominant diffusers in the propagation environment, from the measured temporal and angular power profiles (spatiotemporal representation of the channel impulse response). From the delay, direction of arrival (DOA) and information gain relating to each cluster, its location relative to the base station and the mobile as well as its attenuation law (law in $N_i.\log_{10}$ (length_ ray i)), can be calculated. We then obtain a photograph of the propagation environment considered for the selected measured case.

Figure 2.38 gives an example of the spatiotemporal representation of the complex impulse response measured during a broadband multisensor experiment in a dense environment at 2 GHz and in a small cell context (antenna above roofs) (Laspougeas *et al.* 2000).

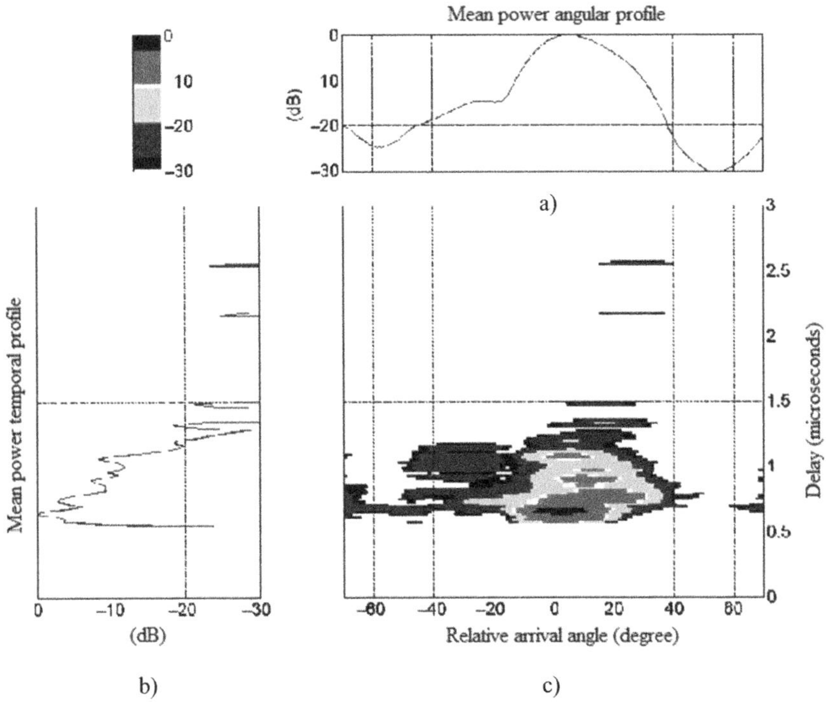

Figure 2.38. *Spatiotemporal representation of the impulse response: a) angular power profile, b) temporal power profile, c) mean spatio-temporal power distribution, the origin of the angles corresponds to the pointing axis of the antenna at the base station*

2.5.5. *Ultra-wideband propagation*

Ultra-wideband (UWB) systems use signals over a very wide frequency band (500 MHz to several GHz) in the band 3.1–10 GHz. Its broad spectral support and its strong temporal resolution power allow them to transmit at very high bit rates, up to several hundred Mbit/s. To assess their performance, attenuation and wideband models have been proposed based on radio channel measurement data.

2.5.5.1. *Attenuation model*

The propagation attenuation is given by the following equation (Recommendation ITU-R P.1791):

$$PL(d) = PL_0(d_0) + 10n\log_{10}\left(\frac{d}{d_0}\right) + X\sigma \ (dB) \qquad [2.137]$$

where:

– $PL_0(d)$ is the propagation loss over a reference distance d_0 (typically $d_0 = 1$ m);

– d is the distance from transmitter to receiver (m) (d > 1 m);

– n is the power loss coefficient;

– Xσ is the standard deviation of log-normal fading (dB); it translates the slow variations of the channel linked to the irregular phenomena of the shadowing effects.

Attenuation $PL_0(d_0)$ in dB is generally modeled using the following equation:

$$PL_0(d_0) = 20\log_{10}\left(\frac{4\pi d_0\sqrt{f_1 . f_2}}{0.3}\right) \qquad [2.138]$$

where f_1 (GHz) and f_2 (GHz) are frequencies at − 10 dB of the UWB spectrum.

In Table 2.15, we show some typical values of the attenuation coefficient and the standard deviation of the fading according to the environment and the different path types (LOS: line-of-sight; NLOS: non-line-of-sight).

Environment	Path type	n	Xσ (dB)
Residential buildings	LOS	1.7	1.5
	NLOS (slightly obstructed)	3.5–5	2.7–4
	NLOS (strongly obstructed)	7	4
Industrial buildings	LOS	1.5	0.3–4
	NLOS (slightly obstructed)	2.1–4	0.19–4
	NLOS (strongly obstructed)	4–7.5	4–4.75

Table 2.15. *Typical values of the attenuation coefficient and the standard deviation of fading*

2.5.5.2. *Impulse response model*

The most commonly used model is that of Saleh and Valenzuela (Salonen *et al.* 1990). It is based on clusters made up of a set of rays with similar properties. Each of the clusters is represented by a decay function (exponential or linear (dB)) as a function of the delay time (Figure 2.39).

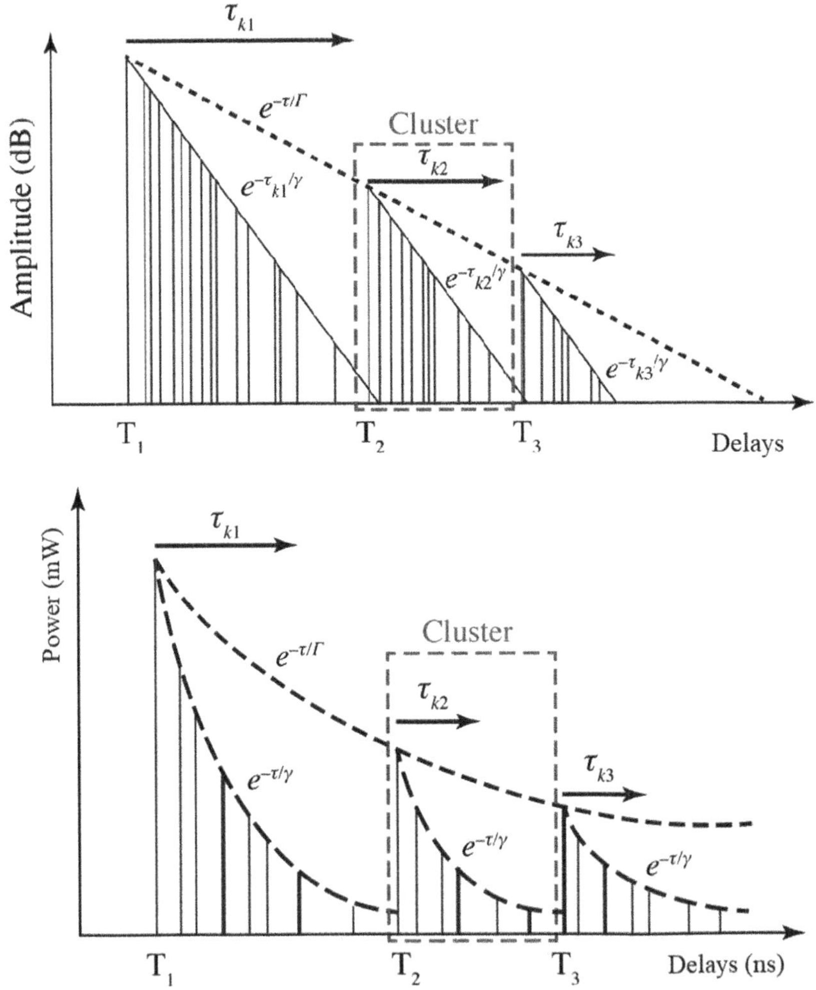

Figure 2.39. *Power profile according to Saleh and Valenzuela formalism*

The impulse response h(τ) has the following general form:

$$h(\tau)=\sum_{k,l}\beta_{kl}\exp\left(j\theta_{kl}\right)\delta\left(t-T_l-\tau_{kl}\right) \qquad [2.139]$$

The amplitude of a path β_{kl} is modeled by a random variable with a Rayleigh distribution and a mean square value respecting an exponential decay:

$$\overline{\beta_{kl}^2}=\overline{\beta_{(0,0)}^2}e^{-\frac{T_l}{\Gamma}}e^{-\frac{\tau_{kl}}{\gamma}} \qquad [2.140]$$

where:

- $\overline{\beta^2{}_{(0,0)}}$ represents the average power of the first path of the first cluster;

- T_l is the arrival time of the first cluster;

- τ_{kl} is the arrival time of the kth path in the first cluster, relative to T_l;

- Γ and γ respectively determine the rate of decrease of the intercluster and intracluster power.

Figure 2.39 shows a representation of the power profile according to the Saleh and Valenzuela formalism (linear (dB) and exponential (mW) representation) (Pagani 2005).

The determination of an impulse response therefore requires the generation of clusters, on the one hand, and that of the different rays within each cluster on the other.

2.5.5.2.1. Generation of clusters

We must first determine the number L of clusters that constitute it. With the arrival of a new cluster being modeled by a Poisson process, the number of clusters can be generated by drawing a random variable L following Poisson's law (Pagani 2005):

$$p_l(L)=\left(\frac{\left(\overline{L}\right)^L\exp\left(-\overline{L}\right)}{L!}\right) \qquad [2.141]$$

where $\overline{L}$ represents the mean number of clusters.

The intercluster duration follows an exponential parameter law. The density of probability of arrival of a new cluster is given by the equation (Saleh and Valenzuela 1987):

$$p(T_l \mid T_{l-1}) = \Lambda e^{-\Lambda(T_l - T_{l-1})} \quad [2.142]$$

where Λ is the cluster arrival rate.

The amplitude of the different clusters follows a power decrease. As a result, the amplitude of the first ray of each cluster is given by the equation:

$$\beta_{1,l}^2 = \beta_{1,1}^2 \left(\frac{T_l}{T_1}\right)^{-\Omega} \quad [2.143]$$

where Ω represents the intercluster power decay coefficient.

The number of rays that can be generated can be infinite with levels becoming infinitely low, therefore it is necessary to choose an observation dynamic D. It will allow the elimination of clusters and the highly attenuated rays. In practice the value of parameter D should not exceed the value of 50 dB.

The different clusters can therefore be generated as follows (Pagani 2005):

1) initialization: the arrival time of the 1st cluster is linked to the transmitter-receiver distance d:

$$l = 1 \text{ and } T_l = d/c \quad [2.144]$$

where c is the speed of light. Let us arbitrarily set the amplitude of this first ray to $\beta_{11} = 1$;

2) generation of a new cluster: to determine the intercluster duration, let us take a random variable ΔT_l according to an exponential parameter Λ law:

$$l = l + 1 \quad [2.145]$$

$$T_l = T_{l-1} + \Delta T_l \quad [2.146]$$

$$\beta_{1,l} = \left(\frac{T_l}{T_1}\right)^{-\frac{\Omega}{2}} \quad [2.147]$$

3) stopping criteria: to validate the last cluster, its amplitude must remain within the observation dynamic. So we check the condition:

$$20\log_{10}\left(\frac{\beta_{k,l}}{\beta_{1,1}}\right) > -D \tag{2.148}$$

so:

$$\tau_{k,l} < T_l\left(10^{\frac{D-G}{10\omega}}\left(\frac{T_l}{T_1}\right)^{-\frac{\Omega}{\omega}} - 1\right) \tag{2.149}$$

2.5.5.2.2. Generation of rays

Within each cluster, the arrival of a ray is modeled by a Poisson process. As in the case of clusters, the density of probability of a new ray is written (Saleh and Valenzuela 1987):

$$p\left(\tau_{kl} \mid \tau_{k-1,l}\right) = \lambda e^{-\lambda\left(\tau_{kl}-\tau_{(k-1)l}\right)} \tag{2.150}$$

where λ is the ray arrival rate in each cluster.

Thus, the generation of the rays inside each of the clusters can be determined in a similar way to that of the clusters. The amplitude of the rays inside each cluster follows a power decay of parameter ω. To consider the power difference observed between the first path of each cluster and the following paths, we use the power ratio G in dB.

The different rays inside each cluster can therefore be generated as follows (Pagani 2005):

1) initialization: the arrival time of the 1st ray corresponds to that calculated for the cluster:

$$\text{k} = 1 \text{ and } \tau_{1,l} = 0 \tag{2.151}$$

The amplitude $\beta_{1,l}$ was previously calculated;

2) generation of a new ray: the inter-ray duration is given by a random variable $\Delta\tau_{k,l}$ generated according to an exponential law of parameter λ, so:

$$k = k + 1$$

$$\tau_{k,l} = \tau_{k-1,l} + \Delta\tau_{k,l}$$

$$\beta_{k,l} = 10^{\frac{-G}{20}} \beta_{1,l} \left(\frac{\tau_{k,l} + T_l}{T_l} \right)^{-\frac{\omega}{2}} \quad [2.152]$$

3) stopping criteria: to validate the last ray, its amplitude must remain within the observation dynamics. So we check the condition:

$$20 \log_{10} \left(\frac{\beta_{k,l}}{\beta_{1,1}} \right) > -D \quad [2.153]$$

so:

$$\tau_{k,l} < T_l \left(10^{\frac{D-G}{10\omega}} \left(\frac{T_l}{T_1} \right)^{-\frac{\Omega}{\omega}} - 1 \right) \quad [2.154]$$

In the event of a stoppage, only the first k-1 rays will be kept ($K_l = k - 1$).

An example of model parameter values (Laspougeas *et al.* 2002) is presented in Table 2.16.

	LOS	LOS	NLOS	NLOS
Cluster	Λ (MHz)	Ω	Λ (MHz)	Ω
	36.5	4.4	24.9	3.9
Ray	Λ (MHz)	ω	Λ (MHz)	ω
	5.95	11.1	6.19	10.2

Table 2.16. *Example of parameter values of the impulse response model*

The models used for standardization are based on this principle. The improvements relate mainly to the shape of the clusters, the statistical laws of the cluster arrival times and the rays inside the clusters.

Other models exist in the literature. Let us quote, for example, the path models (COST 207 1989), models with directional paths (Erceg *et al.* 2004; Baum *et al.* 2005), TRSI-BRAN (BRAIN 2000), 802.11 (Naftali 1997; Jahlavan and Levesque 2005), etc.

The standard 802.15.4a (IEEE 802.15) deals with low speed but large range WPAN (up to industrial environments). The 802.15.3a standard (IEEE 802.15) deals with very high bit rate and short range WPAN (a few meters).

The 802.15.4 low bit rate model is based on the approach of Saleh and Valenzuela (1987). The impulse response has been improved by Molish *et al.* (2004).

The 802.15.3 high bit rate model also based on the Saleh and Valenzuela approach has been improved by Foerster *et al.* (2002). The fading law retained for each path of the model is a normal log law and not a Rayleigh law.

2.6. The propagation of visible and infrared waves in the Earth's atmosphere

2.6.1. *Introduction*

Telecommunications operators are faced with an ever-increasing demand for greater volumes of information to be transmitted (voice, data, images). The increase in frequency of the systems used is an advantage because it offers higher bandwidths and thus allow higher bit rates.

The use of Free Space Optic (FSO) networks, in the range of visible and infrared wavelengths, thus constitute a mode of wireless transmission at high bit rates (several hundreds of Mbits/s), at short and medium range (from a few tens of meters to a few kilometers).

The main applications are wireless telephony, data transmission, computer networks and high definition television.

Several factors condition the revival of this technology: the ease and speed of deployment, the lack of regulation, the low cost of the equipment and the bit rates offered (2 Mbits/s to 10 Gbits/s) (Bouchet *et al.* 2006).

This technology uses low power laser beams ensuring little impact on the environment. These laser beams involve the transmission of an optical signal (visible or infrared) through the Earth's atmosphere.

They interact with the different components (molecules, aerosols) of the propagation medium. This interaction is at the origin of a large number of phenomena such as absorption, diffusion, scintillation. Its only limitation is heavy fog and can cover distances up to a few kilometers (five kilometers in clear air). It is therefore suitable for the construction of networks connecting nearby buildings.

One of the challenges still to overcome is gaining better knowledge of the effects of the atmosphere on the propagation in this frequency spectrum, in order to better optimize the synthesis of broadband wireless communication systems and to evaluate their performance. It is a prerequisite for equipment testing. This aspect will be more specifically described here.

Atmospheric effects relating to propagation, such as molecular and aerosol absorption and diffusion, scintillations due to the variation in the air index under the effect of temperature variations, attenuation by hydrometeors (rain, snow, fog) and lithometeors (dust, sand, etc.) as well as their different models (Kruse and Kim, Bataille, Al Naboulsi, Carbonneau, etc.), are presented and assessed alongside experimental results.

The Meteorological Optical Range (MOR), a parameter used to characterize the transparency of the atmosphere, is defined. Different measuring instruments such as the transmissometer and the scatterometer are also described.

Some potential applications are finally mentioned.

2.6.2. *The propagation of light in the atmosphere*

The characteristic performances of atmospheric optical data transmission networks depend on the medium, the Earth's atmosphere, in which they propagate. This, due to its composition, interacts with the light beam (visible or infrared): molecular and aerosol absorption and diffusion by hydrometeors (rain, snow, fog, etc.), by lithometeors (dust, sand, etc.), and scintillation due to the variation in the air index under the effect of temperature variations.

Atmospheric attenuation is the result of an additive effect of absorption and scattering of light in the visible and infrared bands (IR) by gas molecules and aerosols in the atmosphere. It is described by Beer's law, giving the transmittance as a function of the distance:

$$\tau(d) = \frac{P(d)}{P(0)} = e^{-\sigma d} \quad [2.155]$$

where:

– $\tau(d)$ is the transmittance at distance d from the transmitter;

– $P(d)$ is the signal power at distance d from the transmitter;

– $P(0)$ is the power transmitted;

– σ is the attenuation or the extinction coefficient per unit length.

The attenuation is linked to the transmittance by the following expression:

$$Aff_{dB}(d) = 10\log_{10}\left(\frac{1}{\tau(d)}\right) \quad [2.156]$$

The extinction coefficient σ is the sum of the four following terms:

$$\sigma = \alpha_m + \alpha_n + \beta_m + \beta_n \quad [2.157]$$

where:

– α_m is the molecular absorption coefficient (N_2, O_2, H_2, H_2O, CO_2, O_3, etc.), the reader should refer to the structure and composition of the atmosphere;

– α_n is the absorption coefficient by aerosols (fine solid or liquid particles present in the atmosphere – ice, dust, smoke, etc.);

– β_m is the Rayleigh scattering coefficient resulting from the interaction of light with particles smaller than the wavelength;

– β_n is the Mie scattering coefficient, which appears when the particles encountered are of the same order of magnitude as the wavelength of the transmitted wave.

Absorption dominates in the infrared, whereas it is the scattering which is predominant in the visible and the ultraviolet spectrum.

2.6.2.1. *Molecular absorption*

Molecular absorption results from the interaction between the radiation and the atoms and the molecules of the medium (N_2, O_2, H_2, H_2O, CO_2, O_3, Ar, etc.). It defines different transmission windows in the visible and infrared domain (Figure 2.40).

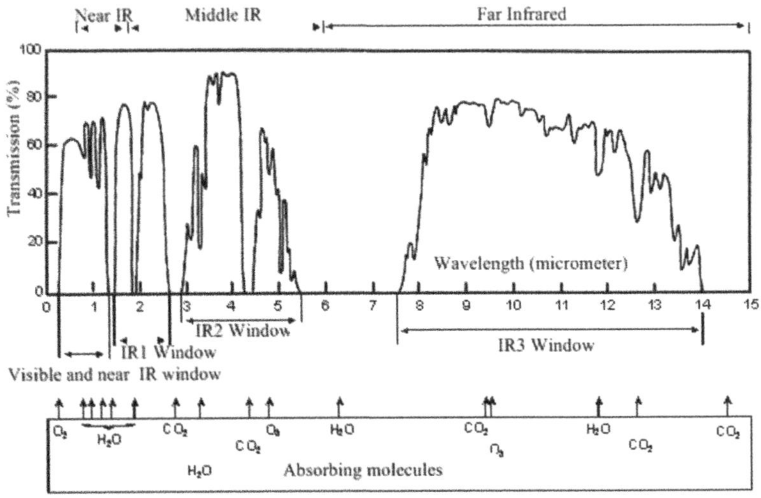

Figure 2.40. *Transmittance of the atmosphere due to molecular absorption*

2.6.2.2. *Molecular scattering*

Molecular scattering results from the interaction of light with particles smaller than the wavelength.

An approximate value of $\beta m(\lambda)$ is given by the following equations:

$$\beta_m(\lambda) = A\lambda^{-4} \qquad [2.158]$$

$$A = 1.09 \times 10^{-3} \frac{P}{P_0} \frac{T_0}{T} \ (\mathrm{km}^{-1}\mu\mathrm{m}^4) \qquad [2.159]$$

where:

– P (mbar) is the atmospheric pressure and $P_0 = 1{,}013$ mbar;

– T (K) is the atmospheric temperature $T_0 = 273.15$ K.

As a result, this scattering is negligible in the infrared. Rayleigh scattering mainly concerns the UV range up to the visible range of the spectrum. This is responsible for the blue color of a clear sky.

2.6.2.3. *Aerosol absorption*

Aerosol absorption results from the interaction between radiation and aerosols, fine particles suspended in the atmosphere (ice, dust, grains of sand, smoke, fog).

The absorption coefficient αn is given by the following equation:

$$\alpha_n(\lambda) = 10^5 \int_0^\infty Q_a\left(\tfrac{2\pi r}{\lambda}, n''\right) \pi r^2 \frac{dN(r)}{dr} dr \qquad [2.160]$$

where:

- $\alpha_n(\lambda)$ is the absorption coefficient by aerosols (km^{-1});
- λ is the wavelength (μm);
- dN(r)/dr is the size distribution of particles per unit volume (in cm^{-4});
- n'' is the imaginary part of index n of the considered aerosol;
- r is the particle radius (cm);
- Q_a (2πr/λ, n'') is the absorption cross section for a given type of aerosol.

The Mie scattering theory (Mie 2008) allows the electromagnetic field diffracted by homogeneous spherical particles to be determined. It makes it possible to evaluate the two physical variables, which are the normalized absorption cross section Q_a and the normalized scattering cross section Q_d. They depend on the size of the particles, their refractive index and the wavelength of the incident radiation. They represent the section of an incident wave normalized by the geometric cross section of the particle (πr^2), such that the absorbed (scattered) power is equal to the power passing through this section.

The refractive index of aerosols depends on their chemical composition. It is complex and depends on the wavelength, denoted n = n' + n'' where n' is related to the scattering power of the particle and n'' relates to the absorbing power of this same molecule.

Note that in the visible and near infrared range, the imaginary part of the refractive index is extremely small and can be disregarded in the calculation of the overall attenuation (extinction). In the far infrared, however, this is not the case.

2.6.2.4. *Aerosol scattering*

This results from the interaction of light with particles (aerosols, hydrometeors, lithometeors) of the same order of magnitude as the wavelength. It is mainly due to the haze and mists in the considered wavelength range.

The scattering coefficient β_n is given by the following equation:

$$\beta_n(\lambda) = 10^5 \int_0^{\infty} Q_d\left(\tfrac{2\pi r}{\lambda}, n'\right) \pi r^2 \frac{dN(r)}{dr} dr \qquad [2.161]$$

where:

– $\beta_n(\lambda)$ is the aerosol scattering coefficient (km^{-1});

– λ is the wavelength (μm);

– dN(r)/dr is the particle size distribution per unit volume (cm^{-4});

– n' is the real part of index n of the aerosol considered;

– r is the particle radius (cm);

– Q_d (2πr/k)λ, n'') is the scattering cross section for a given type of aerosol.

The particle size distribution is generally represented by an analytical function such as the lognormal distribution for aerosols and the modified Gamma distribution for fog (Kruse *et al.* 1962; Bataille 1992; Kim, McArthur and Korevaar 2001; Al Naboulsi, Sizun and De Fornel 2003). The latter is widely used to model the different types of fog and clouds. It is given by the following equation (Shettle and Fenn 1979; Deirmendjian 1969):

$$N(r) = ar^{\alpha} \exp(-br) \qquad [2.162]$$

where:

– N(r) is the number of particles per unit volume and whose radius is between r and r + dr;

– α, a and b are parameters that characterize the particle size distribution.

Atmospheric transmission calculation software such as FASCOD, LOWTRAN and MODTRAN consider two particular types of fog: thick advection fog and convection or moderate radiation fog, which are modeled by the modified Gamma size distribution. The typical parameters are given in Table 2.17 (Clarke 1968; Shettle 1989).

	α	A	b	N	W	r_m	V
Advection fog	3	0.027	0.3	20	0.37	10	130
Radiation fog	6	607.5	3	200	0.02	2	450

Table 2.17. *The different parameters characterizing the particle size distribution in the case of thick advection fog and radiation fog*

In the table:

– N is the total number of water particles per unit volume (nb/cm^3);

– r_m is the modal radius for which the distribution has a maximum (μm);

– W is the liquid water content (g/m^3);

– V is the visibility associated with the type of fog (m).

The Mie scattering theory allows us to express the scattering coefficient Q_d due to aerosols. It is calculated by assuming that the particles are spherical and sufficiently distant from each other so that the field scattered by a particle and arriving at another can be calculated in the far field.

The scattering cross section Q_d is a function which is highly dependent on the size of the aerosol compared to the wavelength. It reaches its maximum (3.8) by a particle radius equal to the wavelength: the scattering is then maximal. Then, when the size of the particles increases, it stabilizes around a value equal to 2. We must therefore expect a very selective function by particles with a radius less than or equal to the wavelength. Clearly, scattering is highly dependent on the wavelength.

The concentration of aerosols, their composition and their size distributions vary greatly over time and in space, hence the difficulty in predicting the attenuation caused by these aerosols. Although their concentration is closely related to optical visibility, there is no clear particle size distribution for a given visibility. Visibility characterizes the transparency of the atmosphere estimated by a human observer. It is measured by the meteorological optical range (MOR).

The scattering coefficient is the most limiting factor from the point of view of the propagation of atmospheric optical waves. It is the most limiting factor for the deployment of atmospheric optical networks. The attenuation can reach 300 dB/km while the attenuation of millimetric waves by rain is only about 10 dB/km.

Different models exist in the literature: that of Kruse and Kim, Bataille and Al Naboulsi.

2.6.3. *The different models*

2.6.3.1. *The Kruse and Kim models*

The attenuation coefficient for optical and near infrared waves up to 2.4 µm is approximated by the following equation:

$$\gamma(\lambda) \simeq \beta_n(\lambda) = \frac{3.912}{V}\left(\frac{\lambda_{nm}}{550}\right)^{-q} \qquad [2.163]$$

where:

– V is the visibility (km);

– λ_{nm} is the wavelength (nm);

– the coefficient q characterizes the particle distribution. It is given by the following equation (Kruse *et al.* 1962):

$$q = \begin{cases} 1.6 & \text{if } V > 50 \text{ km} \\ 1.3 & \text{if } 6 \text{ km} < V < 50 \text{ km} \\ 0.585 V^{1/3} & \text{if } V < 0.5 \text{ km} \end{cases} \qquad [2.164]$$

As a result, the attenuation is a decreasing function of the wavelength.

Recent studies have led to the parameter q being defined as follows (Kim *et al.* 2001):

$$q = \begin{cases} 1.6 & \text{if } V > 50 \text{ km} \\ 1.3 & \text{if } 6 \text{ km} < V < 50 \text{ km} \\ 0.16V + 0.34 & \text{if } 1 \text{ km} < V < 6 \text{ km} \\ V - 0.5 & \text{if } 0.5 \text{ km} < V < 1 \text{ km} \\ 0 & \text{if } V < 0.5 \text{ km} \end{cases} \qquad [2.165]$$

where V is the visibility (km).

As a result, attenuation is a decreasing function of the wavelength when visibility is greater than 500 meters. For lower visibilities, the atmospheric attenuation is independent of the wavelength.

2.6.3.2. *Bataille's model*

Bataille's model (Bataille 1992) calculates molecular and aerosol extinction for six laser lines (0.83, 1.06, 133, 1.54, 3.82 and 10.591 μm) by a polynomial approach on ground-based terrestrial networks. We describe it below.

2.6.3.2.1. Molecular extinction

The linear extinction coefficient σ_m is obtained via a 10-term expression:

$$\sigma_m = -\ln\begin{pmatrix} B_1 + B_2T' + B_3H + B_4T'H + ... \\ ... + B_5T'^2 + B_6H^2 + B_7T'H^2 + ... \\ ... + B_8T'^2\,H + B_9H^3 + B_{10}T'^3 \end{pmatrix} \qquad [2.166]$$

where:

– T' = T(K)/273.15 is the reduced air temperature;

– H is the absolute humidity (g/m^3).

The coefficients B_i (i = 1,10), for the different wavelengths studied are given in the literature (Bataille 1992; Vasseur *et al.* 1997).

2.6.3.2.2. Aerosol extinction

The linear extinction coefficient σ_n is obtained via a 10-term expression:

$$\sigma_n = -\ln\begin{pmatrix} A_1 + A_2H + A_3H^2 + A_4H^x + ... \\ ... + A_5V^{-1/2} + A_6V^{-y} + A_7HV^{-1/2} + ... \\ ... + A_8\left(H/V\right)^y + A_9H^z/V + A_{10}HV^{-1} \end{pmatrix} \qquad [2.167]$$

where:

– V is the visibility (km);

– H is the absolute humidity (g/m^3);

– x, y, and z are real values used to optimize the polynomial for each of the wavelengths studied, with their value being adjusted so that the maximum relative error between FASCOD2 and the polynomial is less than 5%. The coefficients A_i (i = 1.10), for the different waves studied, are given in the literature for two types of aerosol: rural and maritime (Bataille 1992; Cojan and Fontanella 1990).

2.6.3.3. *The Al Naboulsi model*

Al Naboulsi *et al.* (2003) developed simple relationships from FASCOD to assess the attenuation in the wavelength range 690 to 1,550 nm and visibilities ranging from 50 to 1,000 m for two types of fog: advection fog and convection fog.

Advection fog occurs when warm, moist air moves over cold ground. The air in contact with the ground cools and reaches its dew point. Water vapor condensation appears. It appears, more particularly, in spring when there is movement of warm, humid air from the south over snow-covered regions. Attenuation by advection fog is expressed by the following equation:

$$\sigma_{advection} = \frac{0.11478\lambda}{V} \qquad [2.168]$$

where:

– λ is the wavelength (μm);

– V is the visibility.

Radiation or convection fog is caused by cooling of the air mass by night radiation from the ground when conditions are favorable (very low winds, high humidity, clear sky). The soil loses its accumulated heat during the day, and becomes cold. The air cools down on contact, reaches its dew point and the humidity it contains increases. A cloud forms that touches the ground, especially in valleys.

Attenuation by convection fog is expressed by the following equation:

$$\sigma_{convection} = \frac{0.18126\lambda^2 + 0.13709\lambda + 3.7502}{V} \qquad [2.169]$$

where:

– λ is the wavelength (μm);

– V is the visibility.

2.6.3.4. *Attenuation by rain*

Attenuation by rain (dB/km) is generally given by the Carbonneau equation (Carbonneau and Wiseley 1998):

$$Att_{rain} = 1.076R^{0.67} \qquad [2.170]$$

Figure 2.41 shows variations in specific attenuation (dB/m) due to precipitation in the optical and infrared spectrum.

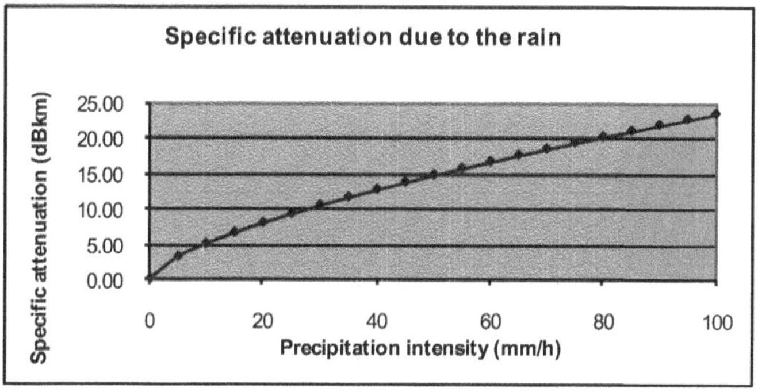

Figure 2.41. *Specific attenuation (dB/km) due to rain in the optical and infrared range*

Recommendation ITU-R P.837 gives the intensity of rain R_p, exceeded for a given percentage of the average year, p, and at a given location.

2.6.3.5. *Attenuation by snow*

Snow attenuation as a function of snowfall rate is given by the following equation:

$$Att_{snow}\left[dB/km\right] = aS^b \qquad [2.171]$$

where:

– Att_{snow} is the snow attenuation (dB/km);

– S is the snowfall rate (mm/h);

– a and b are functions of the wavelength given by the following equations as a function of the wavelength in nanometers (Table 2.18).

	A	b
Wet snow	$0.0001023\lambda_{nm} + 3.7855466$	0.72
Dry snow	$0.0000542\lambda_{nm} + 5.4958776$	1.38

Table 2.18. *Values of coefficients a and b help calculate the snow attenuation (wet and dry)*

The attenuations as a function of the snowfall rate at 1,550 nm are given in the Figures 2.42 and 2.43.

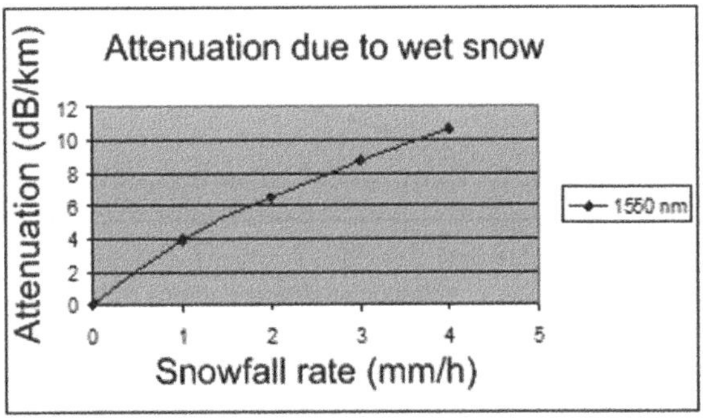

Figure 2.42. *Wet snow: attenuation as a function of precipitation rate at 1,550 nm*

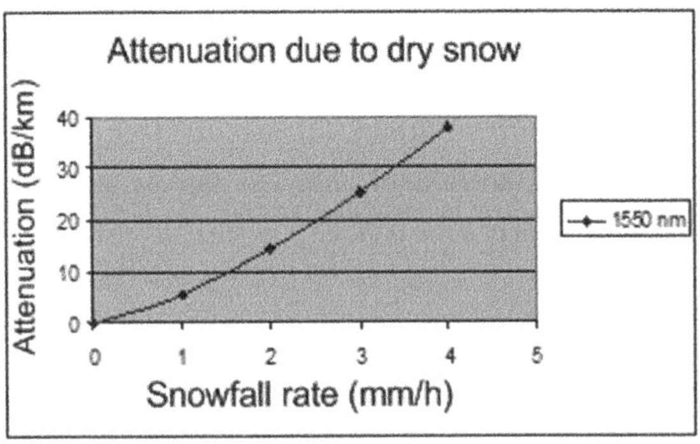

Figure 2.43. *Dry snow: attenuation as a function of precipitation rate at 1,550 nm*

2.6.3.6. *Scintillation*

Under the influence of thermal turbulence, within the propagation medium, we witness the formation of randomly distributed cells, of variable sizes (from 10 cm to 1 km) and of different temperatures. These different cells have different refractive indices, thus causing scattering, multiple paths and variation in the angles of arrival: the received signal fluctuates quickly at frequencies between 0.01 and 200 Hz. The wave front varies in a similar way, causing beam focusing and defocusing. Such fluctuations in the signal are called scintillations.

Figures 2.44 to 2.46 show this effect as well as the variations (amplitude, frequency) in the received signal. When the heterogeneities are large compared to the cross section of the beam, it is deflected (Figure 2.44); when they are small, the beam is broadened (Figure 2.45). When the heterogeneities are different sizes, large and small, scintillations occur (Figure 2.46) (Weichel 1989).

The effect of tropospheric scintillation is generally studied using the logarithm of the amplitude χ [dB] of the "log-amplitude" signal observed, defined as the ratio, in decibels, of its instantaneous amplitude to its average value. The intensity and the speed of the fluctuations (frequency of scintillations) increase with wave frequency. For a plane wave, a weak turbulence and a point receiver, the "log-amplitude" scintillation variance of σ_χ^2 [dB^2] can be expressed by the following equation:

$$\sigma_\chi^2 = 23.17 k^{7/6} C_n^2 L^{11/6} \quad [2.172]$$

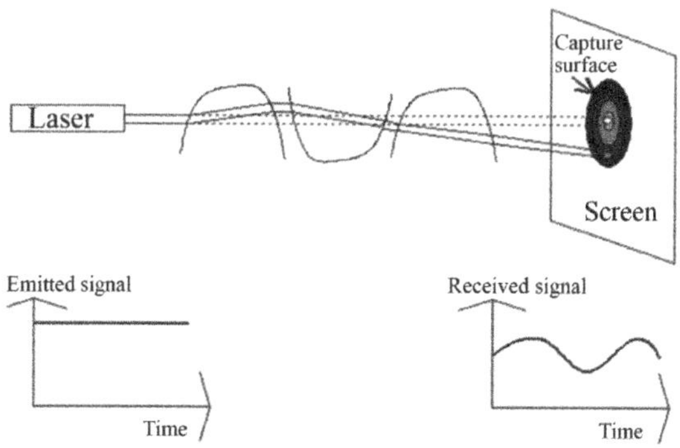

Figure 2.44. *Deviation of the laser beam under the influence of turbulence cells greater than the beam diameter (deviation of the beam)*

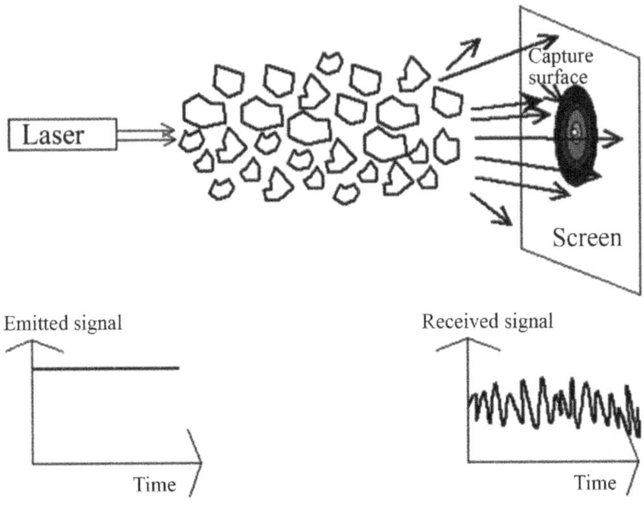

Figure 2.45. *Deviation of the laser beam under the influence of turbulence cells smaller than the beam diameter (beam enlargement)*

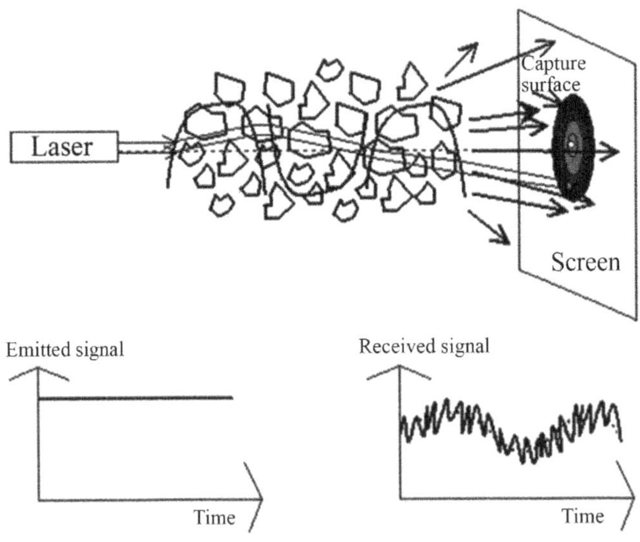

Figure 2.46. *Effects of different heterogeneities and different sizes on the propagation of a laser beam (scintillations)*

where:

– k is the wave number ($2\pi/\lambda$) (m^{-1});

– L is the link length (m);

– C_n^2 is the parameter describing the structure of the refractive index, representing turbulence intensity ($m^{-2/3}$).

The peak to peak amplitude scintillation is $4\sigma_\chi$, and the attenuation linked to the scintillation is $2\sigma_\chi$. For strong turbulences, we observe a saturation of the variance given by the above relationship (Bataille 1992). Note that parameter C_n^2 does not have the same value for millimeter waves and optical waves (Vasseur *et al.* 1997). Millimeter waves are especially sensitive to fluctuations in humidity while in optics, the refractive index is essentially a function of temperature (the contribution of water vapor is negligible). For millimeter waves, we obtain a value of C_n^2 of around 10^{-13} $m^{-2/3}$ which is a medium turbulence (generally $10^{-14} < C_n^2 < 10^{-12}$ for millimeter waves) and for optical waves a value of C_n^2 of around $2\text{x}10^{-15}$ $m^{-2/3}$ which is weak turbulence (typically in optics $10^{-16} < C_n^2 < 10^{-13}$) (Bataille 1992).

Figure 2.47 gives the variation in the attenuation of optical beams with a wavelength of 1.5 µm for different types of turbulence over distances up to 2,000 meters.

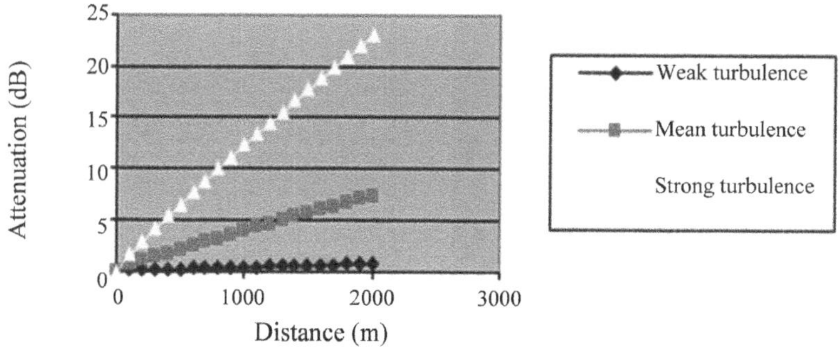

Figure 2.47. *Variation in attenuation linked to the scintillation as a function of distance for different types of turbulence at 1.55 micron*

The international visibility code giving the visible losses (dB/km) for different climatic conditions can be found in Appendix 5 (Kim, McArthur, Korevaar 2001):

– metrological conditions (very clear weather to dense fog);

– precipitation (mm/h: drizzle, rain, thunderstorm);

– visibility (from 50 km to 50 m).

2.6.4. *Experimental results*

In this section, we present some experimental results (Figures 2.48 and 2.49) deduced from attenuation measurements according to the visibility, carried out within the framework of the COST 270 project, in collaboration with the University of Graz (Gebbart *et al.* 2004).

2.6.4.1. *Comparison with the Kruse and Kim models (850 and 950 nm)*

Figures 2.48 and 2.49 show the evolution of the specific attenuation (dB/km) measured at the Turbie site with a light beam at 850 nm and 950 nm as a function of visibility in the presence of fog. The results are compared to the Kruse model (Figure 2.48) and Kim model (Figure 2.49).

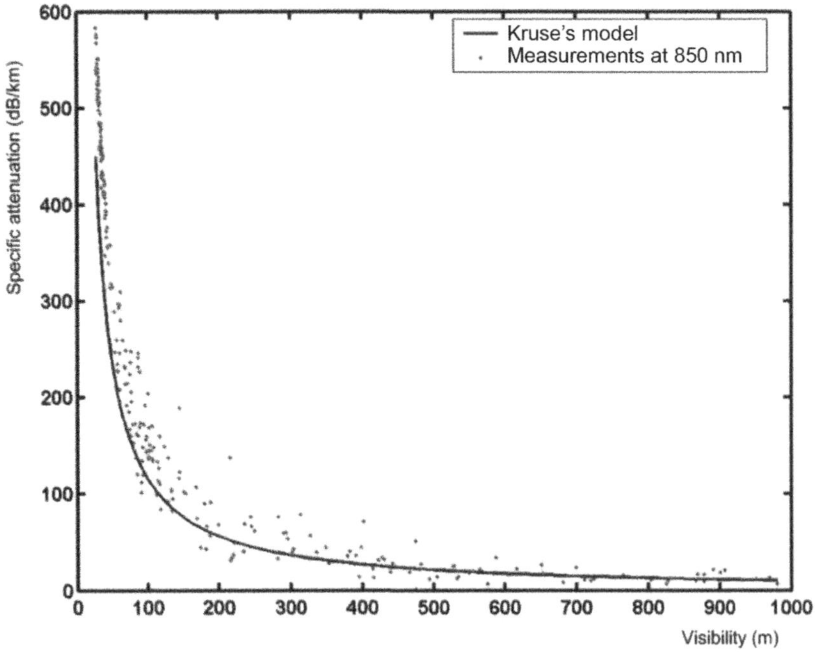

Figure 2.48a. *Variation in specific attenuation at 850 nm as a function of visibility (comparison with the Kruse's model)*

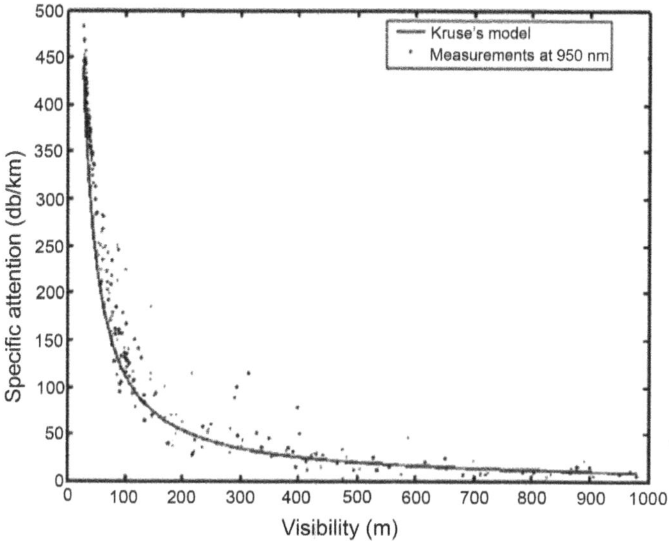

Figure 2.48b. *Variation in specific attenuation at 950 nm as a function of visibility (comparison with the Kruse's model)*

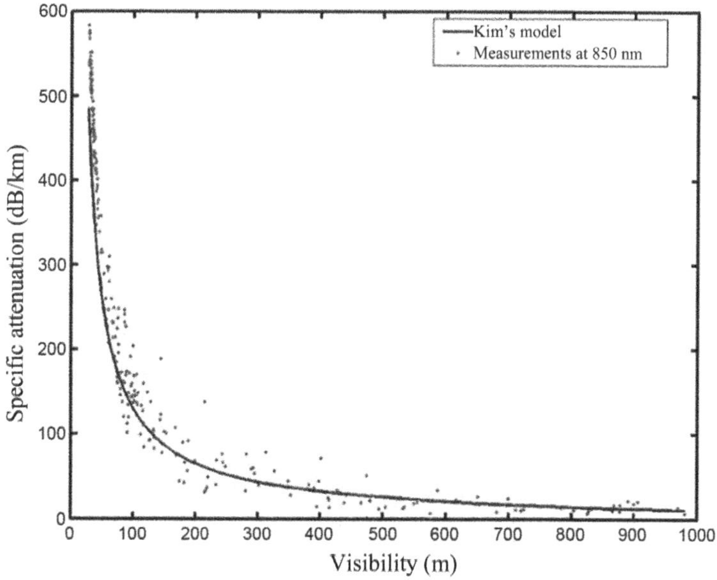

Figure 2.49a. *Variation in specific attenuation at 850 nm as a function of visibility (comparison with Kim's model)*

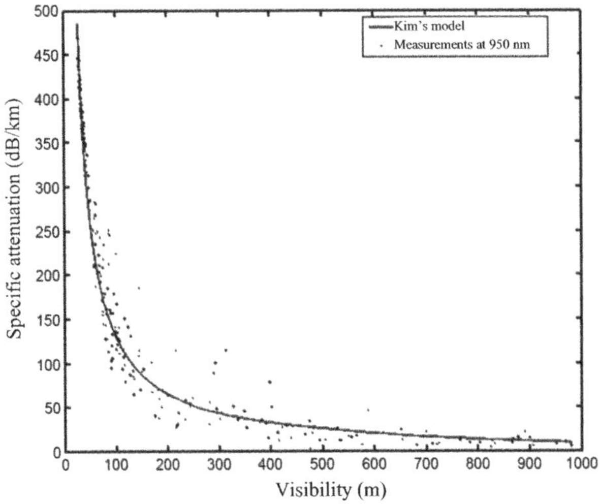

Figure 2.49b. *Variation in specific attenuation at 950 nm as a function of visibility (comparison with Kim's model)*

2.6.4.2. *Comparison with the Al Naboulsi model*

Figures 2.50 and 2.51 show the evolution of the specific attenuation (dB/km) of the light beam at 850 nm measured at the Turbie site as a function of visibility in the presence of fog. The results are compared to the advection and convection model of Al Naboulsi.

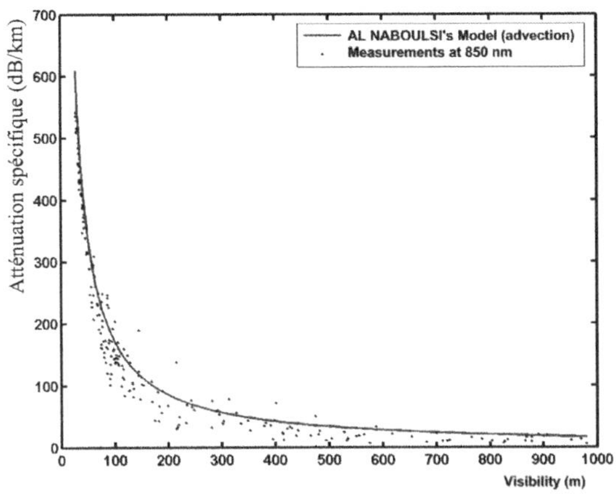

Figure 2.50. *Variation in specific attenuation at 850 nm as a function of visibility (comparison with the Al Naboulsi advection model)*

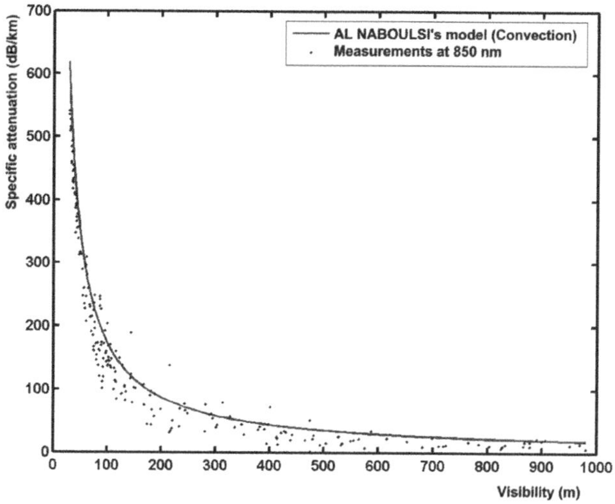

Figure 2.51. *Variation in specific attenuation at 850 nm as a function of visibility (comparison with the Al Naboulsi convection model)*

Comparing the measurements to existing models in the literature reveals a good correlation between the proposed measures and the models. From the analysis of the previous curves, it appears that the Al Naboulsi model, developed from FASCOD, is in excellent agreement with the experimental measurements for low visibilities, whereas the Kruse and Kim models deviate significantly from the measurements.

2.6.5. *Fog and mist*

Fog and mist consist of fine water droplets (<100 μm) suspended in the air. In some very cold countries, there can be ice particles. They are formed by a process of condensation near the ground following a radiative cooling of the earth (especially at night) or the passage of air over cold ground (fog or advection mist).

To distinguish between mist and fog, it is generally accepted that visibility is greater than 1,000 meters in mist while it is less than 1,000 meters in fog. Their compositions and size distributions vary greatly. In general, fog has a lower water content than clouds (< 100 cm^{-3}). It is characterized by visibility, determined by the maximum distance beyond which an object is no longer distinguishable by a human observer. The occurrence of fog is very variable depending on the geographic location: proximity to the sea or a large lake, mountainous area, etc.

Figure 2.52 shows the distribution, over France, of the number of days per year with fog (days during which we observe, if only temporarily, a reduction in visibility to less than one kilometer).

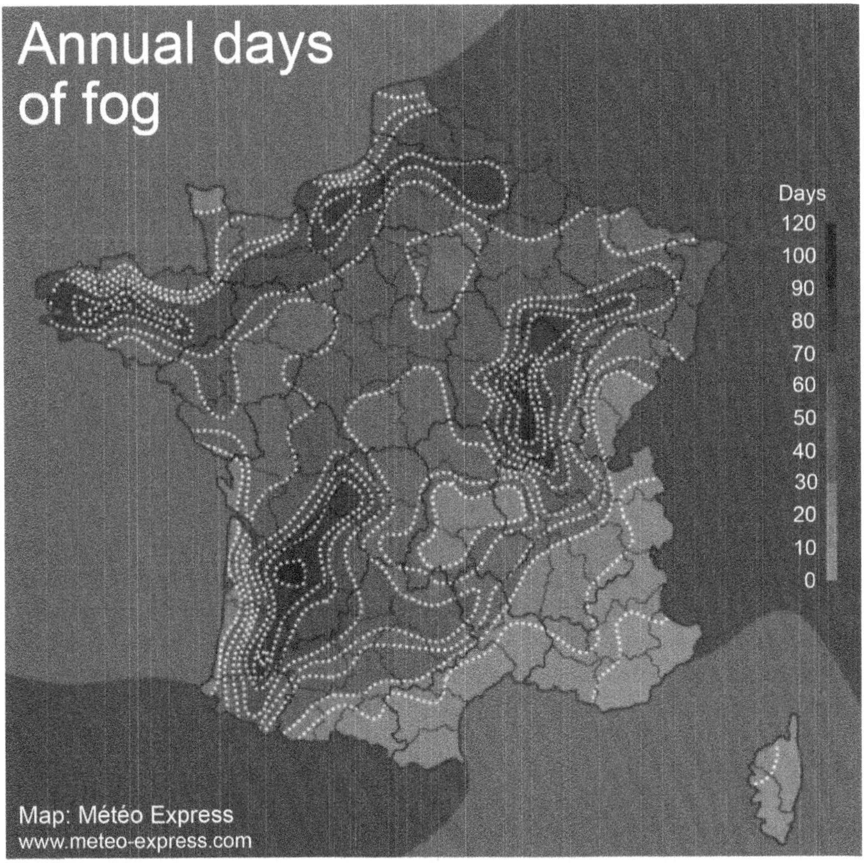

Figure 2.52. *Number of days a year in France with fog (visibility less than 1 km)*

2.6.6. *Sandstorms*

Sandstorms are meteorological phenomena produced by strong winds causing deflation (the release of the finest particles from a sediment) and the transport of sand particles into the atmosphere. They considerably reduce visibility. Figure 2.53 provides an overview of a sandstorm.

Figure 2.53. *Sandstorm (source: Wikipedia)*

2.6.7. *Meteorological optical range*

2.6.7.1. *Visibility*

Originally defined for meteorological purposes, the atmospheric optical range (or visibility) is defined by the length of the path that a beam of parallel light rays must emit in the atmosphere, emanating from an incandescent lamp, at a color temperature of 2,700 K, so that the intensity of the luminous flux is reduced to 0.05 times its initial value. This characterizes the transparency of the atmosphere.

Figure 2.54 gives an example of the variations in the meteorological optical range observed at the Turbie site (06) on June 28, 2004, during a day of low visibility (<10,000 m) and in fog (< 1,000 m).

Visibility is measured using a transmissometer or a scatterometer. The transmissometer is an instrument based on the loss of light intensity from a beam of light rays in the atmosphere, which depends on both absorption and scattering. The scatterometer gives an indication of the visibility in the atmosphere after measuring the scattering by a given volume of a beam of light.

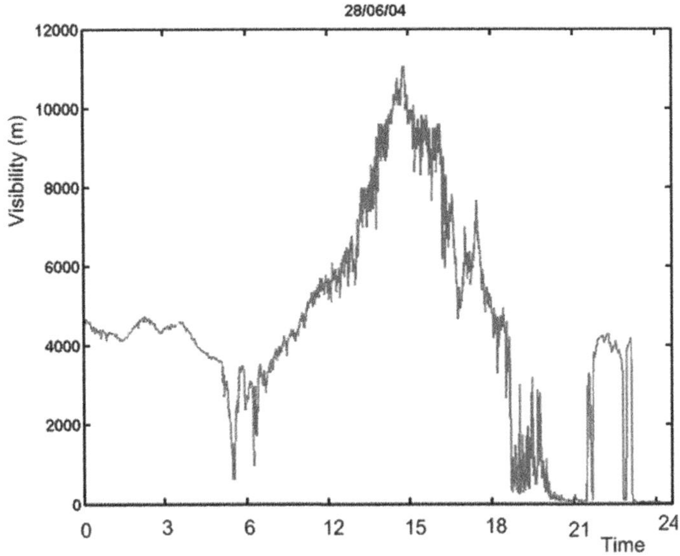

Figure 2.54. *Variations in the MOR observed at the Turbie site on June 28, 2004*

2.6.7.2. *Measuring instruments*

2.6.7.2.1. Transmissometer

The transmissometer method is most commonly used to measure the average extinction coefficient in a horizontal air cylinder placed between a transmitter composed of a light source with constant and modulated flux and a receiver equipped with a photodetector (most often a photodiode located at the focus of a parabolic mirror or lens). The most often used light source is a halogen lamp or xenon light discharge lamp. The modulation of the light source avoids the influence of stray solar light. The current from the photodetector determines the transmission factor, which makes it possible to calculate the extinction coefficient and the MOR.

There are two types of transmissometer (Cojan and Fontanella 1990):

– those whose transmitter and receiver are placed in different boxes and placed at a known distance from each other (Figure 2.55);

– those whose transmitter and receiver are placed in the same box, the light emitted is reflected by a mirror or retro reflector placed at a distance (Figure 2.56).

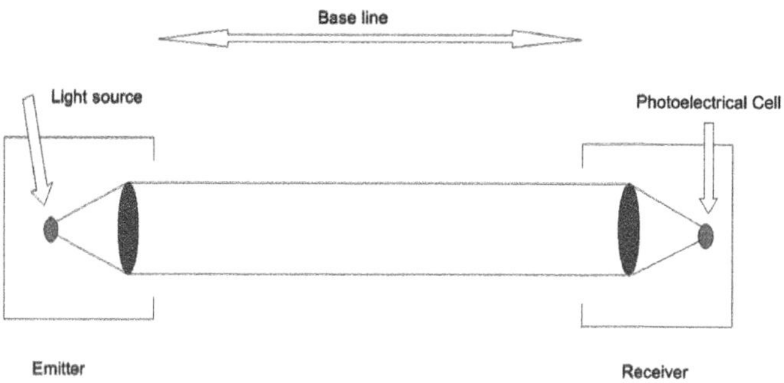

Figure 2.55. *Direct beam transmissometer*

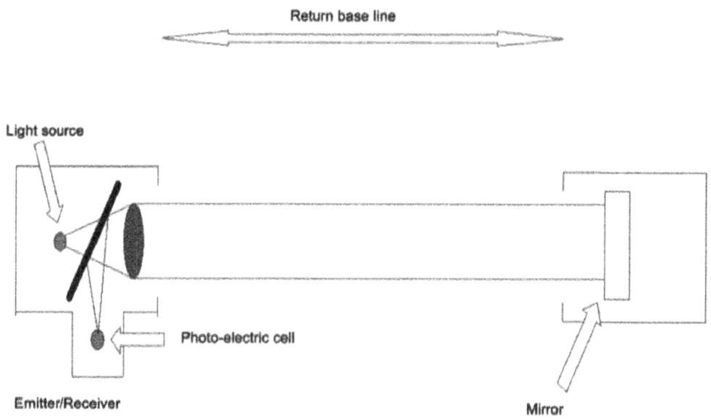

Figure 2.56. *Reflected beam transmissometer*

The distance traveled by light between the transmitter and the receiver is commonly known as the "transmissometer base" and can vary from a few meters to 300 meters.

2.6.7.2.2. Scatterometer

The most practical method to perform this measurement is to focus a light beam on a small volume of air and to determine, by photometric means, the proportion of light scattered in a large enough solid angle and in not preferred directions.

Two types of measurements are used in these instruments (Cojan and Fontanella 1990):

– backscatter (Figure 2.57): the light beam is concentrated in a small volume of air, it is scattered and collected by the photoelectric cell;

– forward scatter: the instruments consist of a transmitter and a receiver whose transmission and reception beams form, between them, an angle of 20 to 50 degrees (Figure 2.58); Other devices place a diaphragm half-way between the transmitter and receiver, or two diaphragms placed close to the transmitter and the receiver.

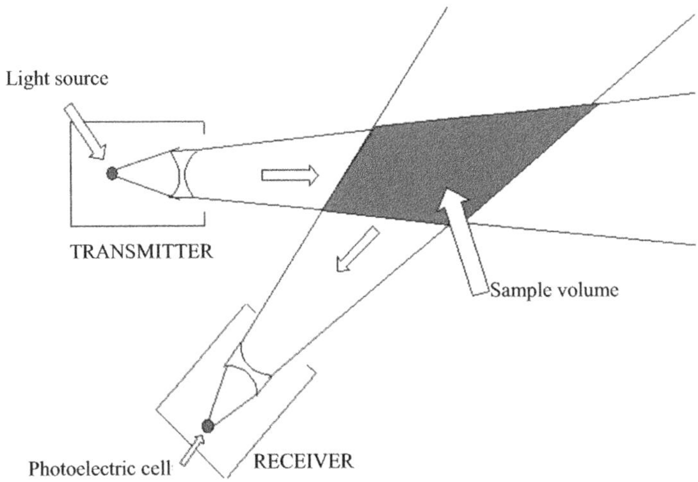

Figure 2.57. *Diagram showing the measurement of visibility by backscatter*

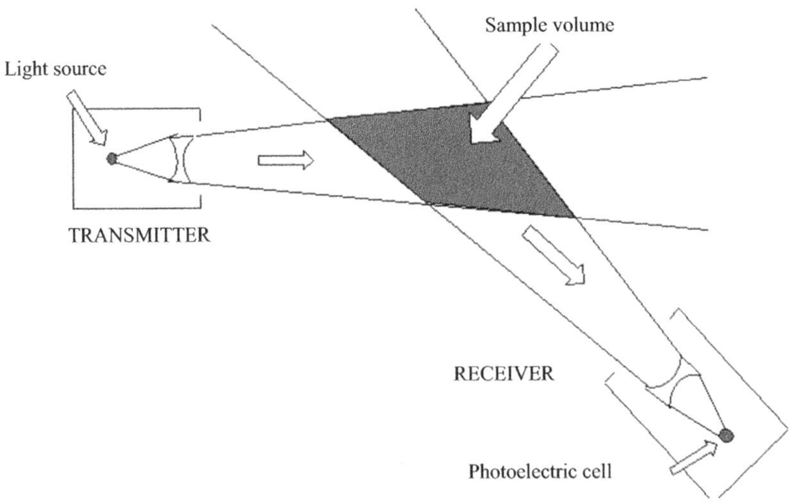

Figure 2.58. *Diagram showing the measurement of visibility by forward scatter*

2.6.8. *Applications*

Atmospheric free-space optical links can be a quick and inexpensive solution in the following situations (Bouchet *et al.* 2006):

– when the user does not own the land on which he wishes to lay his transmission cable (optical fiber, copper cable);

– when obstacles are difficult to cross, such as highways, river crossings, etc.;

– when the link must be operational very quickly and at low cost;

– when the link must be temporarily operational for very specific events (trade fairs, Olympic games, racing, commercial action to meet a customer's need, etc.);

– when the geographic environments are subject to significant climatic disturbances, such as lightning, or saturated in radioelectric networks (factory, airport, etc.);

– when the network must be installed urgently to restore a temporarily destroyed connection (for example, during the September 11, 2001 attack in New York, numerous optical networks were installed to ensure telephone and computer connections);

– when we want to close an optical loop for security reasons (ATM at 155 Mbits/s, WDM at 10 Gbits/s);

– when you want to have high bit rate computer links (Ethernet: 10 Mbits/s, FastEthernet: 100 Mbits/s, Gigabit Ethernet: 1 Gigabits/s, FDDI, etc.) or telephone connections;

– when we want to connect different urban cells, GSM, UMTS, 3G, 4G, 5G, etc.

At the more operational level, we can cite the following applications:

– backhaul in mobile telecommunications: atmospheric optical links are an alternative to cable connections allowing a significant transfer of information between base stations and network infrastructure. It also overcomes radio links, widely used during the deployment of the 2G network (GSM and DCS). With a higher bandwidth, it is suitable for multimedia applications;

– high speed internet connection over the last mile: atmospheric optical links are a practical and economical solution to cover the "last mile" in addition to radioelectric links. Benefiting from the progress made regarding the development of optical fibers, they offer significantly faster speeds and do not require any user licenses;

– secure links: the small sizes of the beams and their invisibility (use of infrared) ensure that atmospheric optical links are concealed. As a result, they are more particularly used in government bodies (defense in particular), and by companies seeking confidentiality. Interception of the beam can only be achieved by placing the same material on the path of the beam, thereby interrupting communication with the user. Information encryption systems are also used to deal with any eventuality;

– high definition television: atmospheric optical links, allowing the transmission of data at high bit rates, are perfectly suited to the transfer of a large number of digital images between buildings, for example, during special temporary events.

2.7. Conclusion

The different aspects of the propagation of photons in the Earth's atmosphere have been presented (molecular and aerosol absorption, molecular and aerosol scattering, attenuation by rain and snow, sandstorms, effects of scintillations). They are key to any good understanding of future communication systems using the wireless approach. Fog appears to be the most detrimental element to atmospheric free-space optical links.

The comparison of experimental data made it possible to validate the models proposed in the literature. These make it possible to control the transmission power levels of future atmospheric optical links while guaranteeing sufficient dynamics that consider the variability in optical propagation conditions.

The experimental links show that atmospheric optical links are a reliable broadband alternative to optical fibers and lead to a better acceptance of this technology in the industry of high-speed telecommunication networks.

In order to better understand the availability of an atmospheric optical link, the reader should refer to the quality of service simulation tools. They help, for a given geographic site, to determine the availability and reliability of a connection according to the system parameters (power, wavelength, characteristics of the material) and climatic and atmospheric parameters. They integrate the various physical phenomena responsible for the breakdown of networks such as attenuations due to ambient light, scintillation, rain, snow, sandstorms and fog (Chabane *et al.* 2005).

This book gives rise to the publication of a study on optical wireless telecommunications (Bouchet 2012) and recommendations by the radiocommunication section of the International Telecommunication Union (ITU-R P.1817 and P.1814) as well as the defense of a doctoral thesis entitled "Contribution to the study of atmospheric optical links: propagation, availability and reliability" (Al Naboulsi 2005).

2.8. Recommendations ITU-R

P.453. The radio refractive index: Its formula and refractivity data [Online]. Available at: https://www.itu.int/rec/R-REC-P.453/en.

P.676. Attenuation by atmospheric gases and related effects [Online]. Available at: https://www.itu.int/rec/R-REC-P.676/en.

P.834. Effects of tropospheric refraction on radiowave propagation [Online]. Available at: https://www.itu.int/rec/R-REC-P.834/ en.

P.837. Characteristics of precipitation for propagation modelling [Online]. Available at: https://www.itu.int/rec/R-REC-P.837/en.

P.1057. Probability distributions relevant to radiowave propagation modelling [Online]. Available at: https://www.itu.int/rec/R-REC-P.1057/en.

P.1238. Propagation data and prediction methods for the planning of indoor radiocommunication systems and radio local area networks in the frequency range 300 MHz to 450 GHz [Online]. Available at: https://www.itu.int/rec/R-REC-P.1238/en.

P.1791. Propagation prediction methods for assessment of the impact of ultra-wideband devices [Online]. Available at: https://www.itu.int/rec/R-REC-P.1791/en.

P.1814. Prediction methods required for the design of terrestrial free-space optical links [Online]. Available at: https://www.itu.int/rec/R-REC-P.1814/en.

P.1817. Propagation data required for the design of terrestrial free-space optical links [Online]. Available at: https://www.itu.int/rec/R-REC-P.1817/en.

2.9. References

Al Naboulsi, M. (2005). Contribution à l'étude des liaisons optiques atmosphériques : propagation, disponibilité et fiabilité. PhD thesis, Université de Dijon.

Al Naboulsi, M., Sizun, H., de Fornel, F. (2003). Fog attenuation prediction for optical and infrared waves. *Optical Engineering*, 43.

Ballabio, E. (1991). Influence of the atmosphere on interference between radio communications systems at frequencies above 1 GHz. Report. European Cooperation in Science and Technology, Brussels.

Barbot, J.P., Levy, A.J., Bic, J.C. (1992). Estimation of fast fading distribution functions. *Proceedings of the URSI Commission F Open Symposium*, Ravenscar, United Kingdom.

Bataille, P. (1992). Analyse du comportement d'un système de télécommunications optique fonctionnant à 0,83 μm dans la basse atmosphère. PhD thesis, Université de Rennes.

Baum, D.S., Dsalo, J., Del Galdo, G., Milojevic, M., Kyösti, P., Hansen, J. (2005). *An Interim Channel Model for Beyond 3G Systems*. VTC, Stockholm.

Bello, P.A. (1963). Characterization of randomly time-variant linear channels. *IEEE, Transactions on Communications Systems*, 11(4), 360–393.

Blanchard, L. (1999). Contribution à l'étude des propagations longues distances en UHF : mécanismes, mesure de champ et brouillage. PhD thesis, Université de Rennes.

Boithias, L. (1983). *Propagation des ondes radioélectriques dans l'environnement terrestre*. Dunod, Paris.

Boithias, L. (1988). *Propagation des ondes radioélectriques*. Techniques de l'Ingénieur, Saint-Denis.

Bouchet, O. (2012). *Wireless Optical Communications*. ISTE Ltd, London.

Bouchet, O., Sizun, H., Boisrobert, C., de Fornel, F., Favennec, P.-N. (2006). *Free-space Optics*. ISTE Ltd, London.

BRAIN (2000). Project technical requirements and identification of necessary enhancements for HIPERLAN type 2, IST-1999-10050, D3.1.

Braun, W.R., Dersch, U. (1991). A physical mobile radio channel. *IEEE: Transactions on Vehicular Technology*, 40(2), 472–482.

Carbonneau, T.H., Wiseley, D.R. (1998). Opportunities and challenges for optical wireless; the competitive advantage of free space telecommunications links in today's crowded market place. In *SPIE Conference on Optical Wireless Communications*. Boston.

Chabane, M., Al Naboulsi, M., Sizun, H., Bouchet, O. (2005). A new quality of service FSO software. In *The European Conference on Propagation and Systems*. Brest.

Chu, T.-S., Greenstein, L.J. (1999). A quantification of link budget differences between the cellular and PCS bands. *IEEE: Transactions on Vehicular Technology*, 48(1), 60–65.

Clarke, R.H. (1968). A statistical theory of mobile-radio reception. *Bell Labs Technical Journal*, 957–1000.

Clay, M.R., Lenham, A.P. (1981). Transmission of electromagnetic radiation in fogs in the 0.53-10.1μm wavelength range. *Applied Optics*, 20(22), 3831–3833.

CODIT (1994). Final propagation models, public deliverable of the RACE 2020. Project CODIT, R2020/TDE/PS/DS/040.

Cojan, Y., Fontanella, J.C. (1990). *Propagation du rayonnement dans l'atmosphère*. Techniques de l'Ingénieur, Saint-Denis.

Correia, L.M. (2001). *Wireless Flexible Personalized Communications*. Wiley & Sons, New York.

COST 207 (1989). Digital land mobile radio communications. Report, Office for Official Publications of the European Communities, Luxembourg.

COST 231 (1999). Evolution of land mobile radio (including personal) communications. Report [Online]. Available at: http://www.lx.it.pt/cost 231/final_report.htm.

COST 255 (1999). Radiowave propagation modelling for Satcom services at Ku-band and above. Report.

Cuinas, I., Sanchez, M.G. (2004). Wide-band measurements of nondeterministic effects on the BRAN indoor radio channel. *IEEE: Transactions on Vehicular Technology*, (53)53.

Deirmendjian, D. (1969). *Electromagnetic Scattering on Spherical Polydispersions*. Elsevier, Amsterdam.

Dersch, U., Zollinger, E. (1994). Physical characteristics of urban microcellular propagation on antennas and propagation. *Electrical and Electronics Engineers Transactions*, (42)11, 1528–1539.

Electronic Communication Committee (2003). The analysis of the coexistence of FWA cells in the 3.4–3.8 GHz band. ECC 33. Technical report.

Erceg, V., Greenstein, L.J., Tjandra, S.Y., Parkoff, S.R., Gupta, A., Kulic, B., Julius, A.A., Bianchi, R. (1999). An empirically based path loss model for wireless channels in suburban environments. *Electrical and Electronics Engineers*, 17(7), 1205–1211.

Erceg, V. *et al.* (2004). IEEE 802.11 Wireless LANs: TGn channel models, IEEE 802.11-03/940R4. Working document.

EURO-COST-231 (2001). Revision 2: Urban transmission loss models for mobile radio in the 900 and 1 800 MHz bands. Final report [Online]. Available at: http://www.lx.it.pt/cost231/final_report.htm.

Failly, M. (1989). Digital land mobile radio communications. Report, COST 207.

Fechtel, S.A., Meyr, H. (1993). Matched filter bound for treillis-coded transmission over frequency-selective fading channels with diversity. *European Transactions on Telecommunications*, (5)3, 109–120.

Foester, J. (2002). Channel modeling sub-committee. Technical report [Online]. Available at: https://www.researchgate.net/publication/244420596_Channel_Modeling_Sub-Committee_Report_Final.

Gebbart, M., Leitgeb, E., Al Naboulsi, M., Sizun, H., de Fornel, F. (2004). Measurements of light attenuation at different wavelengths in dense fog conditions for FSO applications. Report, COST 270.

Gollreiter, R. (1994). Public deliverable of the race 2084. Report, ATDMA project R2084/ESG/CC3/DS/P/029.

Guisnet, B. (1998). La propagation pour les services de mobilité. *L'écho des Recherches*, 170, 15–24.

Gutierrez, J. (2019). IEEE 802.15 WPAN Task Group 4 [Online]. Available at: http://www.ieee802.org/15/pub/TG4.html.

Hari, K.V.S., Sheikh, K.P., Bushue, C. (2001). Interim channel models for G2 MMDS fixed wireless applications [Online]. Available at: http://ieee802.org/16/tg3/contrib/802163c-00_49r2.pdf.

Hata, M. (1980). Empirical formula for propagation loss in land mobile radio services. *IEEE: Transactions on Vehicular Technology*, 29, 317–325.

Jakes, W.C., Reudink, D.O. (1967). Comparison of mobile radio transmission at UHF and X-band. *IEEE: Transactions on Vehicular Technology*, 10–13.

Jouget, M. (1978). *Ondes électromagnétiques 1 : propagation libre*. Dunod, Paris.

Kattenbach, R., Fruchting, H. (1995). Calculation of system and correlation functions for WSSUS channels from wideband measurements. *Frequenz*, 49(3–4), 42–47.

Kim, I.I., McArthur, B., Korevaar, E. (2001). Comparison of laser beam propagation at 785 nm and 1550 nm in fog and haze for optical wireless communications. *SPIE Conference Proceedings*, 42(14), 26–37.

Klepal, M., Mathur, R., McGibney, A., Pesh, D. (2004). Influence of people shadowing on optimal deployment of WLAN access points. *Institute of Electrical and Electronics Engineers*.

Kruse, P.W., McGlauchlin, L.D., McQuistan, R.B. (1962). *Elements of Infrared Technology: Generation, Transmission and Detection*. John Wiley & Sons, New York.

Laspougeas, P., Pajusco, P., Bic, J.C. (2000). Radio propagation in urban small environment at 2 GHz: Experimental spatio-temporal characterization and spatial wideband channel model. *IEEE: Transactions on Vehicular Technology*.

Lavergnat, J., Sylvain, M. (1997). *Propagation des ondes radioélectriques*. Masson, Paris.

Lee, W.C.Y. (1986). *Mobile Communications Engineering*. McGraw-Hill, New York.

Levy, M. (2000). *Parabolic Equation Methods for Electromagnetic Wave Propagation*. The Institution of Engineering and Technology, London.

Liebe, H.J. (1983). Modeling attenuation and phase of radio waves in air at frequencies below 1 000 GHz. *Radio Science*, 16(6), 1183–1199.

Liebe, H.J. (1985). An updated model for millimetre wave propagation in moist air. *Radio Science*, 20(5), 1069–1089.

Liebe, H.J. (1989). MPM – An atmospheric millimeter-wave propagation model. *Journal of Infrared and Millimetre Waves*, 10, 631–650.

Liebe, H.J., Hufford, G.A., Cotton, M.G. (1993). Propagation modelling of moist air and suspended water/ice particles at frequencies below 1,000 GHz. *AGARD*. Palma de Majorca.

Lott, M., Forkel, I. (2001). A multi-wall-and-floor model for indoor radio propagation. *IEEE: VTS 53rd Vehicular Technology Conference*. IEEE, Rhodes.

McNamara, D.A., Pistorius, C.W.I., Malherbe, J.A.G. (1990). *The Uniform Geometrical Theory of Diffraction*. Artech House, London.

Mie, G. (1908). Contributions to the optics of turbid media, particularly of colloidal metal solutions. *Annals of Physics*, 25, 377–445.

Molisch, A.F., Balakrishnan, K., Cassioli, D., Chong, C.C., Emami, S., Fort, A., Karedal, J., Kunisch, J., Schantz, H., Schuster, U., Siwiak, K. (2004). IEEE 802.15.4a channel model – Final report [Online]. Available at: http://www.ieee802.org/15/pub/04/15-04-0662-02-004a-channel-model-final-report-r1.pdf.

Motley, A.J., Keenan, J.M. (1998). Personal communications radio coverage in building at 900 MHz and 1 700 MHz. *Electronics Letters*, 24(12).

Ney, M. (2004). *Bases de l'électromagnétisme*. Techniques de l'Ingénieur, Saint-Denis.

Okumura, Y., Ohmori, E., Kawano, T., Fukua, K. (1968). Field strength and its variability in UHF and VHF land-mobile radio service. *Review of the Electrical Communication Laboratory*, 16(9).

Pagani, P. (2005). Caractérisation et modélisation du canal de propagation radio en contexte Ultra Large Bande. PhD thesis, Institut d'électronique et de télécommunications de Rennes, France.

Pahlavan, K., Levesque, A.H. (2005). *Wireless Information*. John Wiley & Sons, New York.

Parsons, J.D. (1992). *The Mobile Radio Propagation Channel*. Pentech Press Publishers, Stockholm.

Rana, D., Webster A.R., Sylvain, M. (1993). Origin of multipath fading on a microwave link. *Annals of Telecommunications*, 48, 557–566.

Rappaport, T.S. (2009). *Wireless Communications, Principles and Practices*. Dorling Kindersley, London.

Saleh, A.A., Valenzuela, R.A. (1987). A statistical model for indoor multipath propagation. *IEEE Journal on Selected Area in Communications*, 5(2).

Salonen, E. *et al.* (1990). Study of propagation phenomena for low availabilities, ESA/ESTEC Contract 8025/88/NL/PR. Report.

Shen, X.D. (1995). Study of the propagation mechanisms present in trans horizon links. PhD thesis, University of Portsmouth.

Shettle, E.P. (1989). Models of aerosols, clouds and precipitation for atmospheric propagation studies, Atmospheric propagation in the UV, Visible, IR and MM wave region and related systems aspects. *AGARD Conference Proceedings*, 454(15), 1–13.

Shettle, E.P., Fenn, R.W. (1979). Models for the aerosols of the lower atmosphere and the effects of humidity variations on their optical properties. Report, Air Force Geophysical Laboratory, Hanscom.

Sizun, H. (2003). *La propagation des ondes radioélectriques*. Springer, Paris.

Swales, S.C., Beach, M.A., McGrenham, J.P. (1990). The performance enhancement of multibeam adaptative base stations antennas for cellular land mobile radio systems. *IEEE: Transactions on Vehicular Technology*, 39(1), 56–57.

Vasseur, H., Oestges, C., Van der Vorst, A. (1997). Influence de la troposphère sur les liaisons sans fil aux ondes millimétriques et optiques. Propagation électromagnétique du décamétrique à l'angström, Rennes.

Vilar, E., Spillard, C., Rooryck, M., Juy, M. (1988). Observations of troposcatter and anomalous propagation signal levels at 11 GHz on 155 km path over the sea. *Electronics Letters*, 24, 1205–1205.

Visoz, R., Bejjani, E., Kumar, V. (1998). Matched filter bound for equalization and antenna diversity over mobile radio channels. In *IEEE 1998 International Conference on Universal Personal Communications*, Florence.

Weichel, H. (1989). *Laser Beam Propagation in the Atmosphere*. SPIE Press Book, Bellingham.

Appendix 1

Mathematical Formulae[1]

A1.1. Trigonometric transformation equations

If x and y are any two angles, then the following relationships are verified:

$$\begin{cases} \sin(x \pm y) = \sin(x)\cos(y) \pm \cos(x)\sin(y) \\ \cos(x \pm y) = \cos(x)\cos(y) \mp \sin(x)\sin(y) \end{cases}$$

$$\begin{cases} \sin(x) \pm \sin(y) = 2\sin\dfrac{x \pm y}{2}\cos\dfrac{x \mp y}{2} \\ \cos(x) + \cos(y) = 2\cos\dfrac{x+y}{2}\cos\dfrac{x-y}{2} \\ \cos(x) - \cos(y) = -2\sin\dfrac{x+y}{2}\sin\dfrac{x-y}{2} \end{cases} \qquad [A1.1]$$

$$\begin{cases} \sin 2x = 2\sin x\cos x \\ \cos 2x = \cos^2 x - \sin^2 x \\ \tan 2x = 2\sin\dfrac{2\tan x}{1-\tan^2 x} \end{cases} ; \begin{cases} 2\sin^2\dfrac{x}{2} = 1 - \cos x \\ 2\cos^2\dfrac{x}{2} = 1 + \cos x \\ \cos^2(x) + \sin^2(x) = 1 \\ \cos^2(x) = \dfrac{1}{1+\tan^2 x} \end{cases} \qquad [A1.2]$$

For a color version of all figures in this book, see www.iste.co.uk/favennec/electromagnetic1.zip.

1 Bok, J., Hulin-Jung, N. (1979). *Relativité, ondes électromagnétiques.* Hermann, Paris; Bruneaux, J., Saint-Jean, M., Matricon, J. (2002). *Électrostatique et magnétostatique.* Belin, Paris; Amzallag, E., Cipriani, J., Aïm, J.B., Piccioli, N. (2006). *Électrostatique et électrocinétique.* Dunod, Paris.

Electromagnetic Waves 1,
coordinated by Pierre-Noël FAVENNEC. © ISTE Ltd 2020

Moreover, the hyperbolic functions (*hyperbolic sine sinh, hyperbolic cosine cosh and hyperbolic tangent tanh functions*) are given by the following expressions:

$$\sinh x = \frac{e^x - e^{-x}}{2}; \;\; \cosh x = \frac{e^x + e^{-x}}{2}; \;\; \tanh x = \frac{e^x - e^{-x}}{e^x + e^{-x}} \quad [A1.3]$$

The imaginary number i verifies property $i^2 = -1$. For a given angle x, we obtain:

$$e^{ix} = \cos x + i \sin x\,; \;\; \sin x = \frac{e^{ix} - e^{-ix}}{2i}\,; \;\; \cos x = \frac{e^{ix} + e^{-ix}}{2} \quad [A1.4]$$

A1.2. Series developments

If x is a variable, the following series developments are satisfied:

– *binomial development*:

$$(1+x)^n = 1 + nx + \frac{n(n-1)}{2!}x^2 + \frac{n(n-1)(n-2)}{3!}x^3 + \ldots\ldots \quad [A1.5]$$

$$\text{if } x \ll 1:\;\; (1+x)^n \approx 1 + nx \quad [A1.6]$$

– *Maclaurin series* (Colin Maclaurin (1698–1746), Scottish mathematician):

$$e^x = 1 + x + \frac{x^2}{2!} + \frac{x^3}{3!} + \frac{x^4}{4!} + \ldots\ldots;\; \text{if } x \ll 1:\;\; e^x \approx 1 + x \quad [A1.7]$$

$$a^x = e^{x \ln a} = 1 + \frac{(\ln a)}{1!}x + \frac{(\ln a)^2}{2!}x^2 + \frac{(\ln a)^3}{3!}x^3 + \ldots\ldots \quad [A1.8]$$

$$\text{if } x \ll 1:\;\; a^x = 1 + \frac{(\ln a)}{1!}x \quad [A1.9]$$

$$\ln(1+x) = x - \frac{x^2}{2} + \frac{x^3}{3} - \frac{x^4}{4} + \ldots\ldots \quad [A1.10]$$

$$\text{if } x \ll 1:\;\; \ln(1+x) \approx x \quad [A1.11]$$

$$\sin x = x - \frac{x^3}{3!} + \frac{x^5}{5!} - \frac{x^7}{7!} + \ldots\ldots;\ \text{if } x << 1:\ \sin x \approx x \qquad [A1.12]$$

$$\cos x = 1 - \frac{x^2}{2!} + \frac{x^4}{4!} - \frac{x^6}{6!} + \ldots\ldots;\ \text{if } x << 1:\ \cos x \approx 1 \qquad [A1.13]$$

$$\tan x = x + \frac{x^3}{3} + \frac{2x^5}{15} + \frac{17x^7}{315} + \ldots\ldots;\ \text{if } x << 1:\ \tan x \approx x \qquad [A1.14]$$

– *Taylor series* (Brook Taylor (1685–1731), English mathematician): for a given function $f(x)$, the Taylor series development about x_0 is written:

$$f(x) = f(x_0) + (x - x_0)\left(\frac{df}{dx}\right)_{x=x_0} + \frac{(x-x_0)^2}{2!}\left(\frac{d^2 f}{dx^2}\right)_{x=x_0} + \ldots + \frac{(x-x_0)^n}{n!}\left(\frac{d^n f}{dx^n}\right)_{x=x_0} \qquad [A1.15]$$

Appendix 2

Vector Calculations

A2.1. Vectors in coordinate systems

A2.1.1. *Cartesian coordinate systems*

In 3D Cartesian space, a point M is described by its Cartesian coordinates x, y and z in a direct orthonormal trihedron of basic vectors generally denoted $\vec{e}_x$, $\vec{e}_y$ and $\vec{e}_z$ (Figure A2.1).

In the Cartesian coordinate system:

– the elementary volume is: $d\tau = dxdydz$;

– the position vector $\overrightarrow{OM}$ (Figure A2.1) is defined by the relation:

$$\overrightarrow{OM} = x\vec{e}_x + y\vec{e}_y + z\vec{e}_z \qquad [A2.1]$$

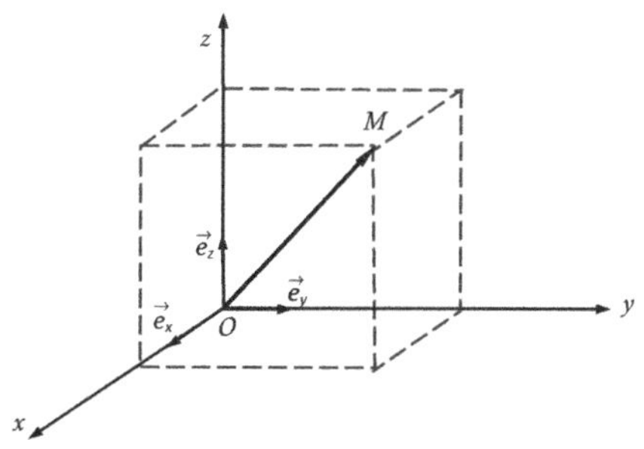

Figure A2.1. *Cartesian coordinate system*

For a color version of all figures in this book, see www.iste.co.uk/favennec/electromagnetic1.zip.

Electromagnetic Waves 1,
coordinated by Pierre-Noël FAVENNEC.

Any vector $\vec{G}$ of components G_x, G_y and G_z is written in Cartesian space dimensions:

$$\vec{G} = G_x \vec{e}_x + G_y \vec{e}_y + G_z \vec{e}_z \qquad \text{[A2.2]}$$

Any displacement vector $\vec{dl} = \overrightarrow{dMM'}$ between two very close points M and M' is written:

$$\vec{dl} = dx\vec{e}_x + dy\vec{e}_y + dz\vec{e}_z \qquad \text{[A2.3]}$$

A2.1.2. *Cylindrical coordinate systems*

If the studied problem presents a cylindrical symmetry compared to an axis of revolution (Oz example), the cylindrical coordinate system is more adapted to the studied problem (case of a rectilinear wire traversed by an electric current where one wants to determine the magnetic field which it creates in a point of space, the point being located on a horizontal plane perpendicular to the wire arranged vertically). In this system, the position of point M is identified by three coordinates (Figure 1.37):

– its distance ρ from the origin O;

– its side z with respect to the xOy plane;

– the azimuth angle θ between the xz plane and the vertical plane containing point M and axis Oz; θ varies between 0 and π.

In the cylindrical coordinate system, a point M (ρ, θ, z) is located in a direct orthonormal local trihedron of basis vectors denoted $\vec{e}_\rho$, $\vec{e}_\theta$ and $\vec{e}_z$. The rotation vectors are such that $\vec{e}_\rho$ is perpendicular to the Oz axis and $\vec{e}_\theta$ is perpendicular to the vertical plane containing point M and the Oz axis. They are oriented as shown in Figure A2.2.

In the cylindrical coordinate system:

– the elementary volume is: $d\tau = \rho d\rho d\theta dz$;

– the position vector $\overrightarrow{OM}$ is given by the relation:

$$\overrightarrow{OM} = \rho\vec{e}_\rho + z\vec{e}_z \qquad \text{[A2.4]}$$

– in the orthonormal basis ($\vec{e}_\rho$, $\vec{e}_\theta$, $\vec{e}_z$), any vector $\vec{G}$ of components (G_ρ, G_θ, G_z) is written:

$$\vec{G} = G_\rho \vec{e}_\rho + G_\theta \vec{e}_\theta + G_z \vec{e}_z \qquad [A2.5]$$

– a displacement vector $d\vec{l} = d\overrightarrow{MM'}$ between two very close points M and M' is written:

$$d\vec{l} = d\rho \vec{e}_\rho + \rho d\theta \vec{e}_\theta + dz \vec{e}_z \qquad [A2.6]$$

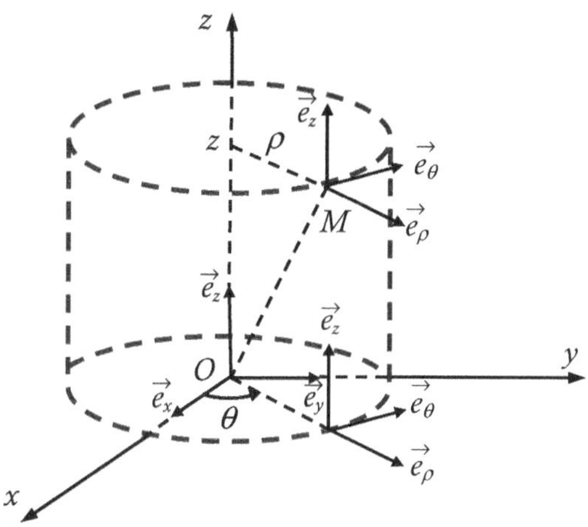

Figure A2.2. *Cylindrical coordinate system*

To understand the origin of the term $\rho d\theta \vec{e}_\theta$ in [A2.6], we derive [A2.4], with respect to time:

$$\frac{d\vec{l}}{dt} = \frac{d\overrightarrow{OM}}{dt} = \frac{d(\rho\vec{e}_\rho)}{dt} + \frac{d(z\vec{e}_z)}{dt} = \frac{d\rho}{dt}\vec{e}_\rho + \frac{\rho d(\vec{e}_\rho)}{dt} + \frac{dz}{dt}\vec{e}_z$$

i.e.:

$$\frac{d\vec{l}}{dt} = \frac{d\rho}{dt}\vec{e}_\rho + \rho\frac{d(\vec{e}_\rho)}{d\theta}\frac{d\theta}{dt} + \frac{dz}{dt}\vec{e}_z = \frac{d\rho}{dt}\vec{e}_\rho + \rho\vec{e}_\theta\frac{d\theta}{dt} + \frac{dz}{dt}\vec{e}_z \qquad [A2.7]$$

By removing dt in the last equality in [A2.7] we find [A2.6].

Note that we used the definition of the derivative of a rotation vector: the derivative of a rotation vector with respect to the angle of rotation is equal to the vector which is directly perpendicular to it by going in the direction of rotation. So considering Figure A2.2, we obtain:

$$\frac{d(\vec{e}_\rho)}{d\theta}=\vec{e}_\theta;\ \frac{d(\vec{e}_\theta)}{d\theta}=-\vec{e}_\rho \qquad [A2.8]$$

A2.1.3. *Spherical coordinate systems*

If the problem studied has a spherical symmetry with respect to the origin of the coordinates, the spherical coordinate system is more appropriate for the problem studied. In this system, the position of point M is identified by three coordinates:

– its distance r from the origin O;

– the azimuth angle φ (or azimuth) between the xz plane and the vertical plane containing point M and the Oz axis; φ varies between 0 and 2π;

– the angle θ (colatitude or zenith) between the $\overrightarrow{OM}$ radius and the Oz *axis*; θ varies between 0 and π.

In the spherical coordinate system, any point M (r, θ, φ) is located in a direct orthonormal local trihedron of basis vectors denoted $\vec{e}_r$, $\vec{e}_\theta$ and $\vec{e}_\varphi$. These three rotation vectors are oriented as shown in Figure A2.3.

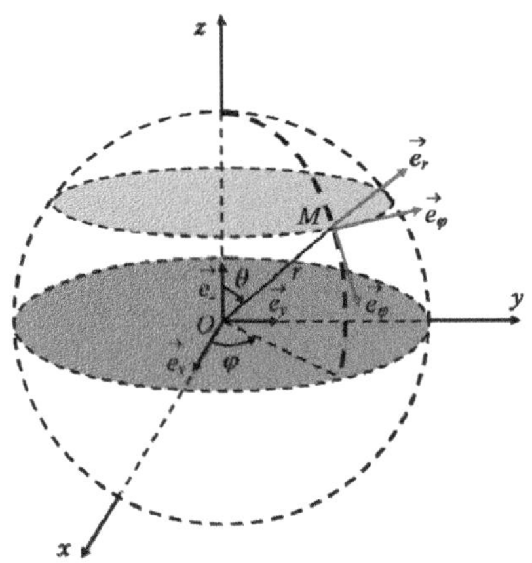

Figure A2.3. *Spherical coordinate systems*

In the spherical coordinate system:

– the elementary volume is: $d\tau = r^2 dr\sin\theta d\theta d\varphi$;

– the position vector $\overrightarrow{OM}$ is given by the relation:

$$\overrightarrow{OM} = r\vec{e}_r \qquad \text{[A2.9]}$$

– in the orthonormal basis ($\vec{e}_r$, $\vec{e}_\theta$, $\vec{e}_\varphi$), any vector $\vec{G}$ of components G_r, G_θ and G_φ is written:

$$\vec{G} = G_\rho\vec{e}_\rho + G_\theta\vec{e}_\theta + G_\varphi\vec{e}_\varphi \qquad \text{[A2.10]}$$

– any displacement vector $\vec{dl} = \overrightarrow{dMM'}$ between two very close points M and M' is written:

$$d\vec{l} = dr\vec{e}_r + rd\theta\vec{e}_\theta + r\sin\theta d\varphi\vec{e}_\varphi \qquad \text{[A2.11]}$$

Cartesian coordinates (x, y, z) are linked to spherical coordinates (r,θ, φ) by the following relations. These relations can be deduced from Figure A2.3:

$$\begin{cases} x = r\sin\theta\cos\varphi \\ y = r\sin\theta\sin\varphi \;;\; OM = r = \sqrt{x^2 + y^2 + z^2} \\ z = r\cos\theta \end{cases} \qquad \text{[A2.12]}$$

A2.1.4. *Laws of orientation in space*

A2.1.4.1. *Notion of direct trihedron*

Let us consider a trihedron formed by the Cartesian basis vectors $\vec{e}_x$, $\vec{e}_y$ and $\vec{e}_z$ (Figure A2.4). By definition, the trihedron ($\vec{e}_x$, $\vec{e}_y$, $\vec{e}_z$) is direct if, for vector $\vec{e}_z$, the rotation of the vector $\vec{e}_x$ on vector $\vec{e}_y$ according to the angle less than 180° occurs in the trigonometric sense.

The trihedron formed from three fingers on the right hand (thumb, index, middle finger) is very practical for determining the direction of Laplace's magnetic force or that of Lorentz's magnetic force.

A2.1.4.2. *Orientation of the surface vector*

For a surface S based on any contour (C), we define a surface vector $\overrightarrow{dS}$ whose modulus is equal to the area dS of the elementary surface, of direction normal to the surface and of direction directly associated with the direction of travel indicated by the corkscrew rule.

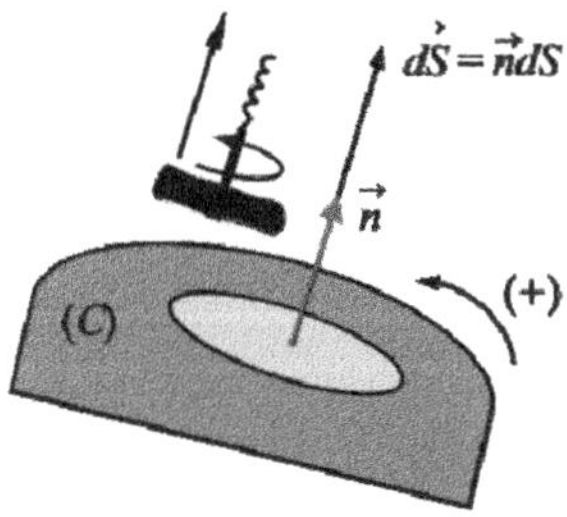

Figure A2.4. *Law of orientation of an elementary surface, corkscrew rule*

If $\vec{n}$ is the vector normal to the surface, then $\overrightarrow{dS} = \vec{n}\,dS$. By arbitrarily choosing a positive direction of travel on the contour (C), the direction of the surface vector is the direction in which the corkscrew advances when it is turned in the chosen positive direction (Figure A2.4).

A2.1.5. *Solid angle*

An angle (in radians) is defined as the ratio of the length l of the arc intersected by the circle, centered at the apex angle, and radius R of this circle, i.e. $\theta = l/R$.

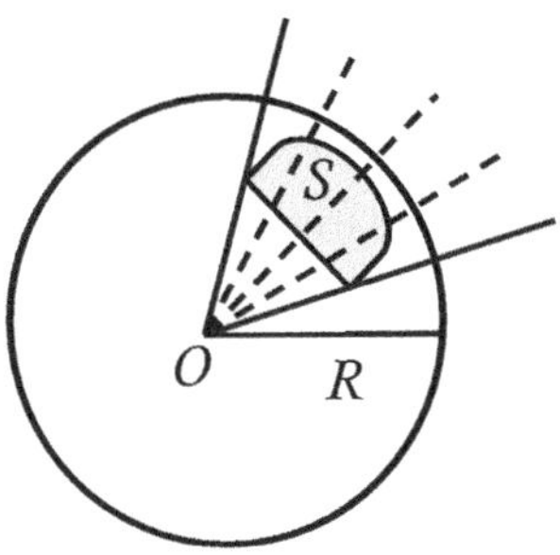

Figure A2.5. *Surface* S *intersected by a cone on a sphere with radius* R

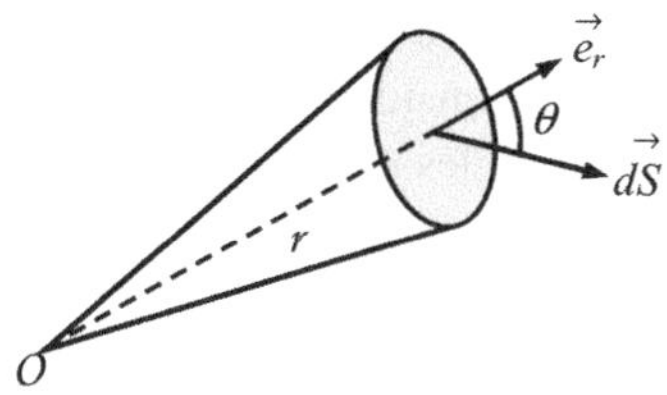

Figure A2.6. *Surface* dS *seen from point* O *at a distance* r

By analogy, we define the solid angle Ω (in *steradians*) of a cone as the ratio of the surface S intersected by this cone on a sphere centered at its apex and the square of the radius of the sphere (Figure A2.5):

$$\Omega = \frac{S}{R^2} \quad \text{[A2.13]}$$

– for any space, $S = \pi R^2 \Rightarrow \Omega = 4\pi$ steradians;

– for a demi-space, $S = 2\pi R^2 \Rightarrow \Omega = 2\pi$ steradians.

Very often, we are led to take an interest in the solid angle which we see from O, an elementary surface dS of small dimensions compared to the distance r which separates the surface of O, and whose normal makes an angle θ with the direction of the unit vector $\vec{e_r}$ (Figure A2.6). This elementary solid angle intervenes for example in the definition of the flow of a vector field through an elementary surface dS. By definition, the elementary solid angle is given by the following relationship:

$$d\Omega = \frac{\overrightarrow{dS} \cdot \vec{e_r}}{r^2} = \frac{dS\cos\theta}{r^2} \quad \text{[A2.14]}$$

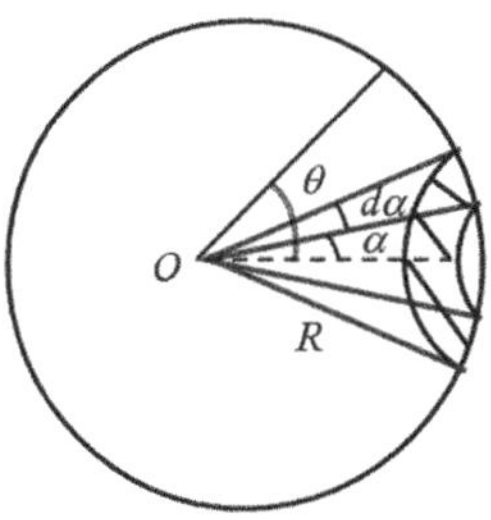

Figure A2.7. *Cones with demi-angles at the apex* α *and* $\alpha + d\alpha$

We express the solid angle Ω of a cone of revolution with apex O and demi-angle at the apex θ. For this, we consider two cones of revolution, with the same apex O, the same axis, and demi-angles α and $\alpha + d\alpha$ (Figure A2.7).

The surface (hatched) cut on the sphere with radius R by the interval between the two cones is given by the relation:

$$dS = 2\pi R \sin\alpha \, R d\alpha = 2\pi R^2 \sin\alpha d\alpha$$

The corresponding elementary solid angle $d\Omega$ is therefore:

$$d\Omega = \frac{dS}{R^2} = 2\pi \sin\alpha d\alpha \qquad \text{[A2.15]}$$

Integrating [A2.15] between the limits 0 and θ, we obtain the solid angle Ω:

$$\Omega = 2\pi \int_0^\theta \sin\alpha d\alpha = 2\pi(1 - \cos\theta) \qquad \text{[A2.16]}$$

Similarly, we find:

– for any space, $\theta = \pi \Rightarrow \Omega = 4\pi$ steradians;

– for a demi-space, $\theta = \pi/2 \Rightarrow \Omega = 2\pi$ steradians.

A2.1.6. *Scalar product of two vectors*

Let us consider two vectors $\vec{G_1}$ and $\vec{G_2}$ forming between them an angle θ (Figure A2.8).

By definition, the scalar product of these vectors is the scalar P such that:

$$P = \overrightarrow{G_1} \cdot \overrightarrow{G_2} = G_1.G_2 \cos\theta \qquad \text{[A2.17]}$$

The scalar product is equal to zero if $\vec{G_1}$ and $\vec{G_2}$ are orthogonal. This is the case of *basis vectors of direct trihedra* of Cartesian, cylindrical and spherical coordinate systems.

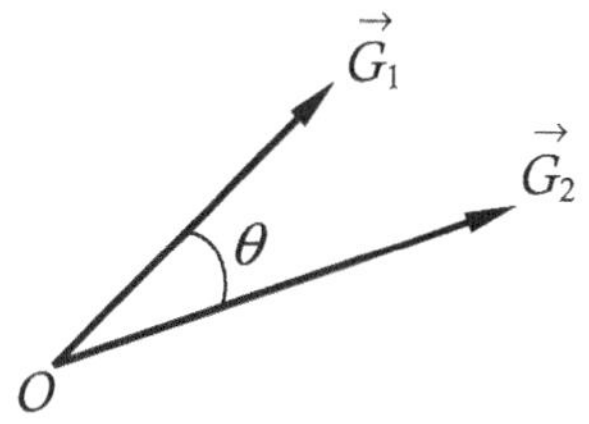

Figure A2.8. *Vectors $\vec{G_1}$ and $\vec{G_2}$ forming angle θ between them*

A2.1.7. *Vector product of two vectors*

Let us consider vectors $\vec{G_1}$ and $\vec{G_2}$ (Figure A2.9). By definition, the vector product of these vectors is the vector $\vec{P}$ such that:

$$\vec{P} = \vec{G_1} \wedge \vec{G_2} \qquad \text{[A2.18]}$$

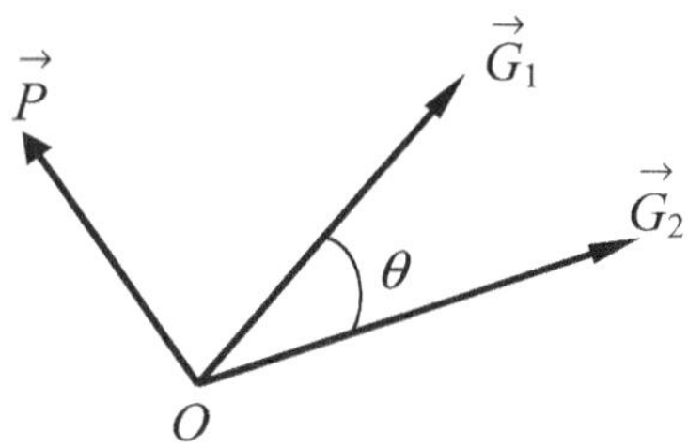

Figure A2.9. *Vector product* P *of two vectors $\vec{G_1}$ and $\vec{G_2}$*

The vector product $\vec{P}$ is perpendicular to the plane formed by the two vectors $\vec{G_1}$ and $\vec{G_2}$, and its sense is such that the trihedron ($\vec{G_1}$, $\vec{G_2}$, $\vec{P}$) is direct (Figure A2.9) and its norm is equal to $G_1G_2|\sin\theta|$. The vector product $\vec{P}$ is null if $\theta = 0$ or π. Moreover, note that the vector product is very important in electromagnetism. It appears in expressions of Laplace's law and the Lorentz law.

A2.1.8. *Field*

A field is the association of a value (of a tensor in the broad sense) of a physical magnitude at each point in space. The space considered may or may not be occupied by a material object.

We distinguish:

– scalar fields (order 0 tensor): at each point in space, we associate a number, for example temperature, pressure, density, etc.;

– vector fields (order 1 tensor): at each point in space we associate a vector, for example the gravity field vector, the electric field vector, the local velocity of a fluid etc.;

– matrix fields (order 2 tensor): at each point, we associate a matrix, for example the elastic deformation tensor, strain tensors.

This notion of a field is most often adapted to the study of continuous environments (mechanics of continuous environments, fluid mechanics) as well as to electromagnetic phenomena. This is essential for the efficient processing of wave phenomena.

We call a field line any "oriented curve" of space which is the tangent at any point to the associated field and oriented in the same direction as the latter. Drawing the field lines establishes the general appearance of the field in a given region of space.

These different fields are represented by functions. When these functions have partial derivatives, we can define other scalars or vectors which remain the same whatever the frame of reference: they are invariants of the scalar field or of the vector field. They provide intrinsic and local properties.

For a *scalar field*, the invariants are the gradient (a vector) and the Laplacian (a scalar).

For a *vector field*, the invariants are the rotational (a vector), the divergence (a scalar) and the vector Laplacian (a vector).

A2.1.9. *Circulation of a vector*

Let us consider that $\vec{G}$ is a vector field and (C) is a circuit whose path is orientated (Figure A2.10).

By definition, the circulation C of the vector field on the circuit is given by:

$$dC = \vec{G} \cdot \vec{dl} \Rightarrow C = \int_{(C)} \vec{G} \cdot \vec{dl} \qquad [A2.19]$$

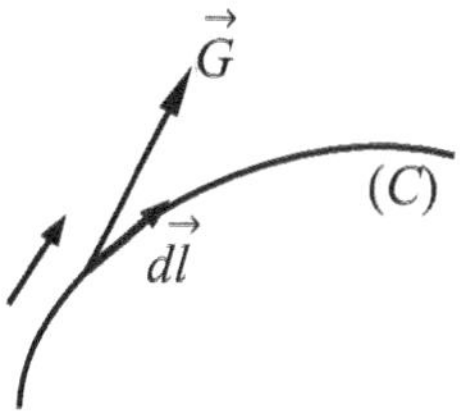

Figure A2.10. *Circulation of a vector on a circuit (C)*

If the considered circuit is closed, then the circulation [A2.19] is written:

$$C = \oint_{(C)} \vec{G} \cdot \vec{dl} \quad \text{[A2.20]}$$

The small circle on the symbol of the integral shows that the integration occurs along the whole circuit.

NOTE.– In the case where the vector field is a force field $\vec{F}$, then the expression [A2.19] gives:

$$C = \int_{(C)} \vec{F} \cdot \vec{dl} = W(\vec{F}) \quad \text{[A2.21]}$$

The result [A2.21] shows that the circulation of a force field on a circuit is nothing other than the work done by the force field on the circuit (for example the work of the weight of a body).

A2.1.10. *Flux of a vector*

Let us consider any vector field $\vec{G}$ and $\vec{dS}$ an elementary surface. The elementary flux of the field $\vec{G}$ leaving the elementary surface is given by the expression:

$$d\Phi = \vec{G} \cdot \vec{dS} \quad \text{[A2.22]}$$

If $\vec{dS} = \vec{n}dS$ is based on a contour (*C*), the total flux leaving the surface *S* is:

$$\Phi = \iint_S \vec{G} \cdot \vec{dS} = \iint_S \vec{G} \cdot \vec{n}dS \quad \text{[A2.23]}$$

The conventions on the orientation of the surface vector were prevoiusly specified (see section A2.1.4.2).

A2.2. Vector operators

A2.2.1. *Gradient operators*

The nabla $\vec{\nabla}$ vector operator is the gradient operator associating a scalar function f and a vector $\overrightarrow{grad} f$ (or $\vec{\nabla} f$) such that for an elementary displacement $\vec{dl}$:

$$df = \vec{\nabla} f \cdot \vec{dl} \qquad \text{[A2.24]}$$

The gradient is an operator which takes an input scalar field and returns a vector. It defines the slope (direction of the greatest variation) of this scalar field at this point.

The nabla operator is involved in many vector calculations in the particular field of electromagnetism which interests us in this book. The expressions for the gradient operator in the various common coordinate systems are as follows.

A2.2.1.1. *Cartesian coordinates*

Scalar function $f(x, y, z)$; direct trihedron ($\vec{e}_x$, $\vec{e}_y$, $\vec{e}_z$):

$$\vec{\nabla} f = \frac{\partial f}{\partial x}\vec{e}_x + \frac{\partial f}{\partial y}\vec{e}_y + \frac{\partial f}{\partial z}\vec{e}_z \qquad \text{[A2.25]}$$

A2.2.1.2. *Cylindrical coordinates*

Scalar function $f(\rho, \theta, z)$; direct trihedron ($\vec{e}_\rho$, $\vec{e}_\theta$, $\vec{e}_z$):

$$\vec{\nabla} f = \frac{\partial f}{\partial \rho}\vec{e}_\rho + \frac{1}{\rho}\frac{\partial f}{\partial \theta}\vec{e}_\theta + \frac{\partial f}{\partial z}\vec{e}_z \qquad \text{[A2.26]}$$

A2.2.1.3. *Spherical coordinates*

Scalar function $f(r, \theta, \varphi)$; direct trihedron ($\vec{e}_r$, $\vec{e}_\theta$, $\vec{e}_\varphi$):

$$\vec{\nabla} f = \frac{\partial f}{\partial r}\vec{e}_r + \frac{1}{r}\frac{\partial f}{\partial \theta}\vec{e}_\theta + \frac{1}{r\sin\theta}\frac{\partial f}{\partial \varphi}\vec{e}_\varphi \qquad \text{[A2.27]}$$

For a path AB along a circuit (C), the circulation of the gradient of a scalar function f is given by the expression:

$$C = \int_{AB} \vec{\nabla} f . \vec{dl} = \int_{A}^{B} df = f(A) - f(B) \qquad \text{[A2.28]}$$

The result [A2.28] shows that the gradient of the scalar function f does not depend on the path followed. In the case of a closed circuit, the circulation of the gradient of a scalar function f is therefore zero:

$$C = f(A) - f(A) = 0$$

A2.2.2. *Divergence operator*

The divergence vector operator div (or even $\vec{\nabla}\cdot$) associates a scalar, denoted div. $\vec{G}$ (or even $\vec{\nabla}\cdot\vec{G}$), to a vector field $\vec{G}$. For a Cartesian coordinate system, the divergence of field $\vec{G}$ is equal to the scalar product of field $\vec{G}$ and the operator nabla $\vec{\nabla}$, i.e.:

$$\text{div}\,\vec{G} = \vec{\nabla}\cdot\vec{G} \qquad \text{[A2.29]}$$

The divergence is an operator that takes an input vector field and returns a scalar.

It represents the flux (per unit volume) of this field (across the surface delimiting a unit of volume at this point).

A *positive divergence* at point M (x,y,z) corresponds to a mainly outgoing flow around this point (divergent field: expansion of a fluid for example).

A *negative divergence* at point M (x,y,z) corresponds to a mainly incoming flow around this point (convergent field: fluid compression for example).

A *zero divergence* at point M (x,y,z) corresponds to incoming and outgoing flows around this point which compensate each other (uniform field: incompressible fluid) or to a swirling flow field.

Like the nabla operator, the divergence operator is involved in many vector calculations in the field of electromagnetism (for example the Green-Ostrogradsky theorem). The expressions of the divergence operator in the case of the various coordinate systems are as follows.

A2.2.2.1. *Cartesian coordinates*

Vector field $\vec{G}$; direct trihedron ($\vec{e_x}$, $\vec{e_y}$, $\vec{e_z}$):

$$\vec{\nabla} \cdot \vec{G} = \frac{\partial G_x}{\partial x} + \frac{\partial G_y}{\partial y} + \frac{\partial G_z}{\partial z} \quad [A2.30]$$

A2.2.2.2. *Cylindrical coordinates*

Vector field $\vec{G}$; direct trihedron ($\vec{e_\rho}$, $\vec{e_\theta}$, $\vec{e_z}$):

$$\vec{\nabla} \cdot \vec{G} = \frac{1}{\rho} \frac{\partial(\rho G_\rho)}{\partial \rho} + \frac{1}{\rho} \frac{\partial G_\theta}{\partial \theta} + \frac{\partial G_z}{\partial z} \quad [A2.31]$$

A2.2.2.3. *Spherical coordinates*

Vector field $\vec{G}$; direct trihedron ($\vec{e_r}$, $\vec{e_\theta}$, $\vec{e_\varphi}$):

$$\vec{\nabla} \cdot \vec{G} = \frac{1}{r^2} \frac{\partial(r^2 G_r)}{\partial r} + \frac{1}{r \sin\theta} \frac{\partial(\sin\theta \, G_\theta)}{\partial \theta} + \frac{1}{r \sin\theta} \frac{\partial G_\varphi}{\partial \varphi} \quad [A2.32]$$

A2.2.3. *Rotation operator*

The rotation vector operator $\overrightarrow{rot}$ (or even $\vec{\nabla} \wedge$) associates vector $\vec{\nabla} \wedge \vec{G}$ to a field of vectors $\vec{G}$. For a Cartesian coordinate system, the rotational of a vector field $\vec{G}$ is equal to the vector product of the field $\vec{G}$ and the nabla operator $\vec{\nabla}$, i.e.:

$$\overrightarrow{rot} \; \vec{G} = \vec{\nabla} \wedge \vec{G} \quad [A2.33]$$

The rotational is a partial derivative operator which associates another vector field to a three-dimensional vector field *G* (*x*,*y*,*z*).

It expresses the tendency of the field lines of a vector field to rotate around a point. The presence of this vector means that this field rotates around this axis. The amplitude of the vector which characterizes this rotation will be greater, the greater the rotation.

Like the nabla and divergence operators, the rotational operator is involved in many vector calculations in the field of electromagnetism (for example, Stokes' theorem). The expressions of the rotational operator in the various coordinate systems are as follows.

A2.2.3.1. *Cartesian coordinates*

Vector field $\vec{G}$ (G_x,G_y,G_z); direct trihedron ($\vec{e_x}$, $\vec{e_y}$, $\vec{e_z}$):

$$\vec{\nabla}\wedge\vec{G}=\left(\frac{\partial G_z}{\partial y}-\frac{\partial G_y}{\partial z}\right)\vec{e_x}+\left(\frac{\partial G_x}{\partial z}-\frac{\partial G_z}{\partial x}\right)\vec{e_y}+\left(\frac{\partial G_y}{\partial x}-\frac{\partial G_x}{\partial y}\right)\vec{e_z} \qquad [A2.34]$$

A2.2.3.2. *Cylindrical coordinates*

Vector field $\vec{G}$ (G_ρ,G_θ,G_z); direct trihedron ($\vec{e_\rho}$, $\vec{e_\theta}$, $\vec{e_z}$):

$$\vec{\nabla}\wedge\vec{G}=\left(\frac{1}{\rho}\frac{\partial G_z}{\partial\theta}-\frac{\partial G_\theta}{\partial z}\right)\vec{e_\rho}+\left(\frac{\partial G_\rho}{\partial z}-\frac{\partial G_z}{\partial\rho}\right)\vec{e_\theta}+\left(\frac{1}{\rho}\frac{\partial(\rho G_\theta)}{\partial\rho}-\frac{\partial G_\rho}{\partial\theta}\right)\vec{e_z} \qquad [A2.35]$$

A2.2.3.3. *Spherical coordinates*

Vector field $\vec{G}$ (G_r,G_θ,G_φ); direct trihedron ($\vec{e_r}$, $\vec{e_\theta}$, $\vec{e_\varphi}$):

$$\vec{\nabla}\wedge\vec{G}=\frac{1}{r\sin\theta}\left(\frac{\partial(\sin\theta G_\varphi)}{\partial\theta}-\frac{\partial G_\theta}{\partial\varphi}\right)\vec{e_r}+\left(\frac{1}{r\sin\theta}\frac{\partial G_r}{\partial\varphi}-\frac{1}{r}\frac{\partial(rG_\varphi)}{\partial r}\right)\vec{e_\theta}+\left(\frac{1}{r}\frac{\partial(rG_\theta)}{\partial r}-\frac{1}{r}\frac{\partial G_r}{\partial\theta}\right)\vec{e_\varphi} \qquad [A2.36]$$

A2.2.4. *Laplacian operator*

The Laplacian operator, or simply the Laplacian, is a vector operator denoted ∇^2 (or even $\vec{\nabla}\cdot\vec{\nabla}=\Delta$). It associates a scalar *f* to another scalar, denoted $\vec{\nabla}\cdot(\vec{\nabla}f)$. By definintion, the Laplacian of a scalar function *f* is equal to the divergence of the gradient of the function *f*, i.e.:

$$\mathrm{div}\cdot(\overrightarrow{grad}\ f)=\vec{\nabla}\cdot(\vec{\nabla}f)=\nabla^2 f=\Delta f \qquad [A2.37]$$

It combines and links the static description of a field (described by its gradient) to the dynamic effects (the divergence) of this field in space and time.

It physically corresponds to a measure of the difference between the value of *f* at any point *M* and the mean value of *f* about point *P*. It can also be interpreted as the local mean curvature. The second derivative (or curvature) thus represents the local deviation of the mean with respect to the value at the point considered.

Like the nabla, divergence and rotational operators, the Laplacian is involved in many vector calculations in the field of electromagnetism (for example the Poisson and Laplace equations). The expressions of the Laplacian in the various coordinate systems are as follows.

A2.2.4.1. *Cartesian coordinates*

Scalar function $f(x, y, z)$; direct trihedron ($\vec{e}_x$, $\vec{e}_y$, $\vec{e}_z$):

$$\nabla^2 f = \frac{\partial^2 G_x}{\partial x^2} + \frac{\partial^2 G_y}{\partial y^2} + \frac{\partial^2 G_z}{\partial z^2} \qquad [A2.38]$$

A2.2.4.2. *Cylindrical coordinates*

Scalar function $f(\rho, \theta, z)$; direct trihedron ($\vec{e}_\rho$, $\vec{e}_\theta$, $\vec{e}_z$):

$$\nabla^2 f = \frac{1}{\rho}\frac{\partial}{\partial \rho}\left(\rho \frac{\partial f}{\partial \rho}\right) + \frac{1}{\rho^2}\frac{\partial^2 f}{\partial \theta^2} + \frac{\partial^2 f}{\partial z^2} \qquad [A2.39]$$

A2.2.4.3. *Spherical coordinates*

Scalar function $f(r, \theta, \varphi)$; direct trihedron ($\vec{e}_r$, $\vec{e}_\theta$, $\vec{e}_\varphi$):

$$\nabla^2 f = \frac{1}{r^2}\frac{\partial}{\partial r}\left(r^2 \frac{\partial f}{\partial r}\right) + \frac{1}{r^2 \sin\theta}\frac{\partial}{\partial \theta}\left(\sin\theta \frac{\partial f}{\partial \theta}\right) + \frac{1}{r^2 \sin^2\theta}\frac{\partial^2 f}{\partial \varphi^2} \qquad [A2.40]$$

A2.2.5. *Relations in vector algebra*

The vector fields $\vec{A}$, $\vec{B}$ and $\vec{C}$ satisfy the following properties:

$$\vec{A} \wedge \vec{B} = -\vec{B} \wedge \vec{A} \qquad [A2.41]$$

$$\vec{A} \cdot (\vec{B} \wedge \vec{C}) = \vec{B} \cdot (\vec{C} \wedge \vec{A}) = \vec{C} \cdot (\vec{A} \wedge \vec{B}) \qquad [A2.42]$$

$$\vec{A} \wedge \vec{B} \wedge \vec{C} = \vec{B} \cdot (\vec{A} \cdot \vec{C}) - \vec{C} \cdot (\vec{A} \cdot \vec{B}) \qquad [A2.43]$$

If f and g are any two scalar functions, then the following properties are also verified:

$$\vec{\nabla}(f+g)=\vec{\nabla}f+\vec{\nabla}g \quad [A2.44]$$

$$\vec{\nabla}(fg)=f\vec{\nabla}g+g\vec{\nabla}f \quad [A2.45]$$

$$\vec{\nabla}\cdot(f\vec{A})=\vec{\nabla}f\cdot\vec{A}+f\vec{\nabla}\cdot\vec{A} \quad [A2.46]$$

$$\vec{\nabla}\cdot(\vec{A}\wedge\vec{B})=\vec{B}\cdot\vec{\nabla}\wedge\vec{A}-\vec{A}\cdot\vec{\nabla}\wedge\vec{B} \quad [A2.47]$$

$$\vec{\nabla}\wedge(f\vec{A})=\vec{\nabla}f\wedge\vec{A}+f\vec{\nabla}\wedge\vec{A} \quad [A2.48]$$

$$\vec{\nabla}\wedge(\vec{\nabla}\wedge\vec{A})=\vec{\nabla}(\vec{\nabla}\cdot\vec{A})-\vec{\nabla}^2\vec{A}=\vec{\nabla}(\vec{\nabla}\cdot\vec{A})-\Delta\vec{A} \quad [A2.49]$$

$$\vec{\nabla}\cdot(\vec{\nabla}f)=\vec{\nabla}^2 f=\Delta f \quad [A2.50]$$

$$\vec{\nabla}\cdot(\vec{\nabla}\wedge\vec{A})=0 \quad ; \quad \vec{\nabla}\wedge(\vec{\nabla}f)=0 \quad [A2.51]$$

A2.3. Integral transform theorems

A2.3.1. *Stokes' theorem*

Stokes' theorem (also known as the rotational theorem) links the flux of a vector field $\vec{G}$ across a closed surface to the rotational of the field:

$$\oint_{(C)}\vec{G}\cdot\vec{dl}=\iint_S(\vec{\nabla}\wedge\vec{G})\cdot\vec{dS} \quad [A2.52]$$

A2.3.2. *Ostrogradsky's theorem*

Ostrogradsky's theorem (also known as the divergence theorem) links the flux of a vector field $\vec{G}$ across a closed surface to the divergence of the field:

$$\oiint_S\vec{G}\cdot\vec{dS}=\iiint_\tau(\vec{\nabla}\cdot\vec{G})\cdot d\tau \quad [A2.53]$$

A2.4. Fundamental relations

The scalar Laplacian is the divergence of the field gradient:

$$div\left(\overrightarrow{grad}\right)=\Delta \qquad [A2.54]$$

The divergence of the rotational of a field is always zero:

$$div\left(\overrightarrow{rot}\right)=0 \qquad [A2.55]$$

The rotational of a field rotational is equal to the gradient of the divergence of this field minus the Laplacian:

$$\overrightarrow{rot}\left(\overrightarrow{rot}\right)=\overrightarrow{grad}\left(div\right)-\Delta \qquad [A2.56]$$

The rotational of the field gradient is zero:

$$\overrightarrow{rot}\left(\overrightarrow{grad}\right)=\vec{0} \qquad [A2.57]$$

Appendix 3

Frequency Spectrum[1]

A3.1. Introduction

Spectrum management at the international level is the responsibility of the ITU-R (International Telecommunication Union – Radiocommunication Sector), headquartered in Geneva. It regularly publishes a frequency allocation table whose role is to share the spectrum according to its use (fixed, mobile, amateur, radio astronomy, etc.) and geographic location worldwide. It is integrated into the Radio Regulations (RR).

Here we provide an overview by frequency range, emphasizing for each, the atmospheric and terrestrial influence, the system considerations and the associated services).

A3.2. The different frequency ranges

A3.2.1. *ELF waves (frequency less than 3 kHz)*

– Wavelength (λ): above 100 km.

– Atmospheric influence: the ionosphere forms the upper limit of the propagation waveguide.

– Terrestrial influence: the surface of the Earth constitutes the lower limit of the propagation waveguide.

– System consideration:

- need for very long antennas (linked to the wavelength);

For a color version of all figures in this book, see www.iste.co.uk/favennec/electromagnetic1.zip.

1. Hall, M.P.M. (1996). Overview of radiowave propagation. In *Radiowave Propagation*, Hall, M.P.M., Barclay, L.W., Hewitt, M.T. (eds). Institution of Electrical Engineers, London.

- very slow information flow;

- sensitivity to noise: antenna and atmospheric (distant stormy discharges, local electrical discharges).

– Associated services:

- submarine communications (short and long distances);

- subterranean communications;

- ultra-sensitive communications on a global scale;

- internal remote sensing of the Earth's subsoil.

A3.2.2. *VLF waves (3–30 kHz)*

– Wavelength(λ): 10–100 km (myriametric waves).

– Atmospheric influence:

- the ionospheric D layer forms the upper limit of the propagation waveguide.

– Terrestrial influence:

- the surface of the Earth constitutes the lower limit of the propagation waveguide;

– System consideration;

- need to raise very long antennas (100 m is only a small fraction of the wavelength);

- difficulty in constructing directional antennas;

- very slow flow of information.

– Associated services:

- world maritime telegraphy;

- long-distance fixed services;

- navigation system (OMEGA);

- standard time diffusion;

- subterranean communications;

- submarine communications;

- geological surveillance.

A3.2.3. *LF waves (30–300 kHz)*

– Wavelength(λ): 1–10 km (kilometric waves).

– Atmospheric influence:

- sky wave in the propagation waveguide formed by the Earth (lower limit) and the ionospheric D layer (upper limit) for frequencies below 100 kHz;

- sky wave and ground wave at frequencies above 100 kHz;

– Terrestrial influence: ground wave is influenced by the Earth's curvature.

– System consideration:

- need to raise very long antennas (100 m is only a small fraction of the wavelength);

- difficulty in building directional antennas.

– Associated services:

- long distance maritime communications;

- fixed services over long distances;

- broadcasting;

- radio navigation.

A3.2.4. *MF waves (300–3,000 kHz)*

– Wavelength(λ): 100–1,000 m (hectometric waves).

– Atmospheric influence:

- ground wave for short distances and lowest frequencies;

- sky wave for long distances and the highest frequencies. The night field is higher even for short distances.

– Terrestrial influence:

- reflection of the wave on the ground.

– System consideration:

- possibility of using directional antennas (multiple element antennas, L or T inverted antennas);

– Associated services:

- broadcasting;

- radio navigation;

- mobile communications (terrestrial, maritime, aeronautical);

- fixed services.

A3.2.5. *HF waves (3–30 MHz)*

– Wavelength(λ): 10–100 m (decametric waves).

– Atmospheric influence:

- ground wave for short distances particularly above the sea;

- sky wave for long distances (beyond the first hop: 1E or 1F).

– Terrestrial influence:

- reflection and scattering.

– System consideration:

- use of log-periodic antenna network (vertical or horizontal);

- use of vertical whip antenna;

- use of horizontal dipole antenna.

– Associated services:

- point-to-point fixed services;

- mobile services (aeronautical, terrestrial, maritime);

- long distance broadcasting.

A3.2.6. *VHF waves (30–300 MHz)*

– Wavelength(λ): 1–10 m (metric waves).

– Atmospheric influence:

- refraction and reflection due to irregularities in the refractive index permitting trans-horizon links;

- ionospheric scattering (on the sporadic layer E in particular) allowing trans-horizon links;

- Faraday rotation and ionospheric scintillation on Earth-space links.

– Terrestrial influence:

- mask effects (hills, mountains);

- diffraction inside valleys;

- reflection on lakes, the sea and flat terrain causing multi-paths effects.

– System consideration:

- use of multi-element dipole antennas (yagi antenna);

- use of slot and helicoidal antenna.

– Associated services:

- broadcasting (up to 100 km);

- mobile communications (terrestrial, aeronautical, maritime);

- mobile telephone;

- aeronautical radio navigation beacon.

A3.2.7. *UHF waves (300–3,000 MHz)*

– Wavelength(λ): 10–100 cm (decimetric waves).

– Atmospheric influence:

- atmospheric refraction;

- reflection on the atmospheric layers at the lowest frequencies;

- ducting effect at higher frequencies;

- fluctuations in the refractive index;

- forescatter beyond the horizon at frequencies above 500 MHz.

– Terrestrial influence:

- mask effects (hills, buildings).

– System consideration:

- use of multi-element dipole antennas (yagi antenna);
- possible use of broadband systems;
- use of reflector antennas at the highest frequencies.

– Associated services:

- television;
- aeronautical navigation;
- radars;
- point to point fixed services;
- mobile communications;
- telemetry;
- cellular communications.

A3.2.8. *SHF waves (3–30 GHz)*

– Wavelength(λ): 1–10 cm (centimetric waves).

– Atmospheric influence:

- attenuation by hydrometeors (rain, snow, hail);
- refraction;
- duct propagation;
- scintillation due to fluctuations in the refractive index.

– Terrestrial influence:

- diffraction by buildings;
- mask effect (buildings, etc.);
- reflection and scattering by the environment (buildings, trees, etc.).

– System consideration:

- use of satellite dishes or horns;
- use of waveguide;
- possibility of using several channels on each carrier.

– Associated services:

- fixed services (voice, multi-channel television);
- radar;
- fixed Earth-satellite link;
- mobile services;
- satellite telemetry.

A3.2.9. *EH waves (30–300 GHz)*

– Wavelength(λ): 1–10 mm (millimetric waves).

– Atmospheric influence:

- significant attenuation by hydrometeors (rain, snow, hail);
- attenuation by clouds, mist, dust, smoke, etc.;
- scattering by hydrometeors (rain, snow, hail);
- scintillations due to fluctuations in the refractive index;
- absorption by oxygen and water vapor.

– Terrestrial influence:

- mask effect (buildings, trees).

– System consideration:

- use of small parabolic antennas;

– Associated services:

- short line-of-sight communications (fixed or mobile);
- satellite remote sensing.

A3.2.10. *Sub-EHF waves (300–3,000 GHz)*

– Wavelength(λ): 0.1–1 mm (sub-millimetric waves).

– Atmospheric influence:

- significant attenuation by meteors (rain, snow, hail, clouds, dust, smoke, etc.);

- appearance of gradient of the refractive index (mirage);
- scintillations due to fluctuations in the refractive index;
- absorption by atmospheric gases (oxygen, water vapor, etc.).

– Terrestrial influence:

- mask effect (trees).

– System consideration:

- use of reflector or lens antennas.

– Associated services:

- short range line-of-sight communications;
- remote sensing.

A3.2.11. *Infrared waves (3–430 THz) and light waves (430–860 THz)*

– Wavelength(λ): 0.1 mm–0.4 μm (infrared and light waves).

– Atmospheric influence:

- significant attenuation by meteors (rain, snow, hail, clouds, dust, smoke, etc.);
- appearance of gradient of the refractive index (mirage);
- scintillations due to fluctuations in the refractive index;
- absorption by atmospheric gases (carbon dioxin, etc.).

– Terrestrial influence:

- mask effect (small objects).

– System consideration:

- use of reflector or lens antennas;
- use of laser.

– Associated services:

- short path communications;
- indoor communications.

Appendix 4

The Decibel

A4.1. Introduction

The decibel (dB) is one tenth of a bel, named after Alexander Graham Bell, inventor of the telephone in 1876. It is a logarithmic unit of measure for the ratio between two powers, two voltages, two pressures, etc. It is used more particularly in the fields of telecommunications, electronics, acoustics, etc. It does not belong to the international system (SI) of units.

A4.2. Definition

It is defined in two different ways depending on whether we consider voltages (V) or powers (P).

In the case of voltages, the variable X in dB is defined by 20 times the logarithmic base 10 of the ratio of the output voltage (V_{out}) to the input voltage (V_{in}):

$$X_{dB} = 20\log_{10}\left(V_{out} / V_{in}\right) \qquad [A4.1]$$

In the case of power, the variable X in dB is defined by 10 times the logarithmic base 10 of the ratio of the output power (P_{out}) to the input power (P_{in}):

$$X_{dB} = 10\log_{10}\left(P_{out} / P_{in}\right) \qquad [A4.2]$$

We will use the term amplification or attenuation depending on whether the variable X_{dB} is positive or negative.

For a color version of all figures in this book, see www.iste.co.uk/favennec/electromagnetic1.zip.

Electromagnetic Waves 1,
coordinated by Pierre-Noël FAVENNEC.

A4.3. The different variants

Different variants exist (dB, dBm, dBW, dBi, dBc, dBμV, dBμVm^{-1}) depending on the references taken by the input voltages and the input powers:

– dB: the reference power is the input power:

$$X_{dB} = 10\log_{10}\left(P_{out} / P_{in}\right) \quad [A4.3]$$

– dBm: the reference power (the input power) is equal to one milliwatt (1 mW):

$$X_{dB} = 10\log_{10}\left(P_{out} / 1mW\right) \quad [A4.4]$$

– dBW: the reference power (the input power) is equal to one watt (1 W):

$$X_{dB} = 10\log_{10}\left(P_{out} / 1W\right) \quad [A4.5]$$

– dBi: here we refer to the gain of an antenna compared to an isotropic aerial. An isotropic aerial is an antenna that emits the same amount of energy in all directions. In reality, such an antenna does not exist:

$$X_{dB} = 10\log_{10}\left(P_{antenna} / P_{isotropicantenna}\right) \quad [A4.6]$$

– dBd: here we refer to the gain of an antenna compared to a half-wave dipole aerial. This aerial has a physical reality:

$$X_{dB} = 10\log_{10}\left(P_{antenna} / P_{dipoleantenna}\right) \quad [A4.7]$$

NOTE.– Catalogs do not generally specify whether they refer to antenna gains in dBi or in dBd. Yet the difference is important. A 10 dBd gain antenna has a gain of 12.15 dBi. We note that there is a great temptation to display gains in dBi rather than gains in dBd without specifying the unit used.

– dBc: the reference power (the input power) is equal to the power of the carrier (P_c) where "c" is the carrier:

$$X_{dB} = 10\log_{10}\left(P_{out} / P_c\right) \quad [A4.8]$$

– dBμV: the reference voltage (the input voltage) is equal to one μVolt (1 μV):

$$X_{dB} = 20\log_{10}\left(Voltage_{out} / 1\mu V\right) \quad [A4.9]$$

– dBμVm^{-1}: the reference field (the input field) is equal to 1 μVm^{-1}:

$$X_{dB} = 10\log_{10}\left(Field_{out} / 1\mu Vm^{-1}\right) \quad [A4.10]$$

A4.4. Decibel operations

As a reminder, we have the following relationships on logarithms:

$$\log_{10}(ab) = \log_{10}(a) + \log_{10}(b) \quad [A4.11]$$

$$\log_{10}(a / b) = \log_{10}(a) - \log_{10}(b) \quad [A4.12]$$

$$\log_{10}\left(a^b\right) = b\log_{10}(a) \quad [A4.13]$$

Also if:

$$A_{dB} = 10\log_{10}(P_{out} / P_{in}) = 10\log_{10}(R) \quad [A4.14]$$

then:

$$R = 10^{\frac{A}{10}} \quad [A4.15]$$

NOTE.– The advantage of using the dB scale is that the gains (amplifications) or attenuations are added up instead of multiplied.

A4.5. Correlation table

R	R (scientific)	A (dB)
0.00001	10^{-5}	– 5
0.0001	10^{-4}	– 4
0.001	10^{-3}	– 3
0.01	10^{-2}	– 2
0.1	10^{-1}	– 1

R	R (scientific)	A (dB)
1	1	0
10	10^{+1}	1
100	10^{+2}	2
1,000	10^{+3}	3
10,000	10^{+4}	4

A4.6. Particular values

Let us consider some particular ratio values, namely 0.1 (1/10), 0.5 (1/2), 2 and 10, giving:

$$10\log_{10}(1/10) = -10 \quad [A4.16]$$

$$10\log_{10}(1/2) = -3 \quad [A4.17]$$

$$10\log_{10}(2) = 3 \quad [A4.18]$$

$$10\log_{10}(10) = 10 \quad [A4.19]$$

Each time we divide by 10 in the current scale, we subtract 10 in the decibel scale.

Whenever we divide by 2 in the current scale, we subtract 3 in the decibel scale.

Appendix 5

The International Visibility Code

Weather condition	Precipitation		mm/h	Visibility (m)	Attenuation (db/km)
Dense fog				0	
				50	315
Thick fog				200	75
Moderate fog				500	28.9
Light fog		Thunderstorm	100	770	18.3
Very light fog				1,000	13.8
	Snow	Heavy rain	25	1,900	6.9
				2,000	6.6
Light mist		Moderate rain	12.5	2,800	4.6
				4,000	3.1
Very light mist		Light rain	2.5	5,900	2
				10,000	1.1
Clear weather		Drizzle	0.25	18,100	0.6
				20,000	0.54
Very clear weather				23,000	0.47
				50,000	0.19

For a color version of all figures in this book, see www.iste.co.uk/favennec/electromagnetic1.zip.

Electromagnetic Waves 1,
coordinated by Pierre-Noël FAVENNEC.

List of Acronyms and Constants

Acronyms

AR *Axial Ratio*

BER *Bit Error Rate*

CAD *Computer-Aided Design*

CEM *Compatibilité Électromagnétique (Electromagnetic Compatibility (EMC))*

CP *Circular Polarization*

CPW *Coplanar Waveguide*

DCS *Digital Cellular System*

DRA *Dielectric Resonant Antenna*

EIRP *Effective Isotropic Radiated Power*

EM *Electromagnetic*

EMC *Electromagnetic Compatibility*

ERP *Effective Radiated Power*

GPS *Global Positioning System*

GSM *Global System for Mobile Communications*

Electromagnetic Waves 1,
coordinated by Pierre-Noël FAVENNEC. © ISTE Ltd 2020

HPBW *Half-Power Beamwidth*

ISI *Inter-Symbol Interference*

Constants

Boltzmann constant	$k = 1.3805 \times 10^{-23}$ joules/degree
Speed of light in vacuum	$c = 2.998 \times 10^{8}$ m.s^{-1}
Vacuum impedance	$\eta 0 \approx 377$
Vacuum permeability	$\mu 0 = 4\pi \times 10^{-7}$ H/m $= 1/\varepsilon_0.c^2$
Vacuum permittivity	$\varepsilon 0 = 8.85419 \times 10^{-12}$ F/m

List of Authors

Jean-Pierre BLOT
EUROSATCOM
Les Ulis
France

Pierre-Noël FAVENNEC
ArmorScience
Lannion
France

Ibrahima SAKHO
University of Thiès
Senegal

Hervé SIZUN
ArmorScience
Lannion
France

Index

A

absorption
  aerosol, 211
  molecular, 210, 232
Ampère (law), 25, 55
angle
  of arrival, 219
  solid, 9
    elementary, 9, 249, 250
antennas, 129, 135, 147, 148, 155, 160, 163, 167, 168, 170, 176, 178, 182, 261–263, 265–268
anti-node, 84
attenuation, 128, 129, 132, 134, 135, 140, 142, 146, 148–150, 152, 156, 161, 163–165, 175–180, 184, 188, 189, 191, 192, 195, 199–201, 208, 211, 214, 215, 221, 232
  free space, 135, 146, 179, 180
axes
  electric, 95, 96
  principle, 96

B, C

bandwidth
  coherence, 142, 146, 196
  correlation, 146
Biot and Savart
  biography, 23
  law, 22, 23, 28, 29, 103
circulation
  of the field
    electrostatic, 5, 6
    magnetic, 25, 33, 104, 105, 107, 108, 116
  of the magnetic excitation vector, 104, 105, 116
coefficient
  absorption
    by aerosols, 209, 211
    extinction, 209, 215, 228
    molecular, 209
complex notation, 81, 84
conservation of energy
  electromagnetic, 64, 69

continuous charge distribution, 6, 10, 51
Coulomb
  biography, 5
  gauge, 26
current
  displacement, 54, 55
  electric, 120, 123, 124
  induced, 34
  magnetization, 98–100, 102–104

D

de Broglie, 73
debye (Dy), 12
delay
  interval, 142, 144, 145, 197
  spread, 144, 146
  window, 142, 145, 196
density
  electromagnetic energy, 65
  magnetization current, 98
  surface density of polarization, 93
  volume charge, 10, 18, 64, 66, 69, 73, 79, 90, 112
depolarization, 131
diamagnetism, 109
dielectric
  homogeneous, 86, 95
  non-homogeneous, 95
dipole
  electrostatic
    definition, 12
    equipotential surfaces, 12, 13
    field lines, 12, 13
  magnetic, 30–32
divergence, 121, 122, , 252, 255, 257, 260
  operator, 58, 100, 255, 256
Doppler spectrum, 139, 197
ducts
  high, 161
  surface, 161

E

Earth's atmosphere, 147, 149, 162, 166, 207, 208, 232
electrical conductivity, 120
electromagnetic, 119, 121
  definition, 35
  invariance, 58
electromagnetism (fundamental equations), 57, 121
electromotive, 41
electrostatic, 2–8, 10, 15, 16, 25, 31, 48, 49, 50, 51
  fundamental laws, 16
equation(s)
  continuity, 111
  Helmholtz, 124–126
  Maxwell, 121, 124, 129
    in a material medium, 86, 115, 117
    in a polarized and magnetic material medium, 114
    in a polarized dielectric medium, 111, 112
    in a vacuum, 1, 23, 57, 79, 85, 112, 113
  Maxwell-Ampère, 56, 65, 79, 103, 114, 116
  Maxwell-Faraday, 56, 65, 93, 116
  Maxwell-flux, 106, 115, 116
  Maxwell-Gauss, 56, 79, 112, 116
  propagation, 64, 70

F

Faraday (biography), 34
ferrimagnetism, 110

ferromagnetism, 110
field
    electric, 119–124, 126, 127, 129, 130, 152
field lines
    electric, 122
        definition, 35, 40
flux of the electrostatic field, 9
force
    Coulomb, 2, 4
    electromagnetic, 35, 40, 42, 58, 66
    electrostatic, 3
    Laplace, 12, 17, 22
    Lorentz, 17, 21, 35
Free Space Optic, 207, 214, 231, 232
Frensel ellipsoids, 133
frequency, 124, 126, 129, 132, 134, 135, 139–142, 146–150, 152, 154, 156, 160, 163, 167, 169, 171, 175–179, 181, 182, 184, 185, 187, 196, 197, 200, 207, 208, 219, 261

G

gauge fixing, 27, 60
Gauss
    biography, 10
    theorem
        integral form, 16
        local form, 16
geographic databases, 147, 173, 178, 185
Global System for Mobile Communications (GSM), 147, 172, 178, 198, 231
gradient, 121, 159–162, 167, 171, 252, 254, 257, 260, 268
    refractive index, 128
Green (biography), 11
guide layers, 128, 171

I

impendance, 127
impulse response, 140–146, 192, 194, 196–200, 202, 203, 206, 207
induction
    electrical, 119
    electromagnetic
        evidence, 33
        first law, 44
        magnetic, 119
        second law, 45
interferences, 133, 138, 144

L

Laplace
    biography, 12
    equation, 11
    law, 17
Laplacian, 34, 121, 252
law
    Coulomb's, 49, 50
    Faraday, 46
    Lorentz's, 17
    Maxwell-Faraday, 45
Lenz
    biography, 34
    law, 1, 34, 42, 46
lines of a field
    electrostatic, 8, 49
    magnetic, 122
links
    cross-horizontal, 171
Lorentz
    gauge, 29
    induction, 18, 46, 47

M

magnetic, 23, 29, 119, 122–124, 126, 127, 129, 148

definition, 35
permeability, 109, 120, 152
susceptibility, 109, 110
magnetization, 86, 103
magnetostatics, 17
dielectric permittivity matrix, 95, 96
fundamental laws, 32, 33
Maxwell (biography), 58
mean delay, 142, 143, 196
medium
dielectric linear, 95–97
magnetic, 98, 103, 105, 107
meteorological optical range (MOR), 208, 213, 227
model
broadband, 196
Motley Keenan, 184
OKUMARA-HATA, 175, 178
Walfish-Ikegami, 178, 179
moment
dipolar, 12, 15, 30, 86
magnetic, 30, 98

N, O

Neumann
biography, 47
induction, 47
node, 84, 85
Ostrogradsky
biography, 11
theorem, 10

P

paramagnetism, 110
paths
diffracted, 132
diffused, 132
direct, 132
multiple, 128, 131, 134, 136, 137, 139, 140, 142, 171, 196, 265
reflected, 132
transmitted, 132
penetration depth, 129
perfect metal, 79, 82
permittivity, 120–122, 129, 152, 168, 185
dielectric
absolute, 95
of the vacuum, 9, 95
principle, 96
relative, 95, 97
phase difference, 129
plane
nodal, 84, 85
polarization, 129
planning tools, 183
Poisson
biography, 12
equation, 11
law, 16
polarization, 129–132, 150, 152, 153, 163, 169
charges, 87, 89, 90
elliptical, 77
potential
circulation, 29
difference, 42, 132, 133, 135, 148, 162–166, 179–181, 186, 194, 196, 237
definition, 26
electrostatic, 4–7, 11, 16, 48
retarded, 62
vector, 26–29, 31–33, 39, 40, 46, 47, 52, 53, 56, 59–64, 69, 70, 73–79, 98, 99, 102
Poynting
identity, 64, 66, 68

principle
charge conservation, 18–20, 55, 64
superposition, 13, 17, 58, 72, 88, 98
product
scalar, 66, 75
vector, 28, 39, 43
propagation, 119, 125–129, 131–136, 140–142, 144, 146–148, 152, 154, 159, 161–163, 166, 167, 171–173, 175, 177–179, 182–187, 194, 196–199, 201, 207, 208, 213, 219, 220, 232, 235–238

R

radioelectric
engineering, 182, 183, 194
visibility, 148, 162
radius of curvature, 160
reflection, 132, 148, 151–155, 157, 161, 162, 167, 170, 171, 186, 194, 196, 219, 264
diffuse, 151, 153
specular, 151, 152, 154
refraction, 148, 157, 161, 162, 167, 171, 172, 220
of the electric field, 91
refractive index, 120, 128, 129, 149, 157, 159–162, 166–168, 171, 173, 211, 221, 264, 265, 266, 267, 268
rotational, 121, 123, 252, 256, 260

S, T

scattering, 148, 154, 163, 166–170, 196, 208, 210–213, 227, 229, 230, 232, 263–266
selectivity, 134, 139, 140, 142, 169, 196, 197
Stokes
biography, 6
theorem, 6, 25, 29, 43, 44, 93, 104, 107
surface of separation
vacuum-dielectric, 91, 93–95
vacuum-magnetic medium, 105–109
theory
geometric theory of diffraction, 163
Mie scattering, 213
transmittance, 209
transpolarization, 131

U, V

ultra- wideband propagation, 200, 237
Universal Mobile Telecommunication Systems (UMTS), 147, 172, 231
variability, 134, 136, 139, 140, 159, 196, 232
vector
current density
definition, 18
flow of, 105
surface magnetization, 101, 102
volume magnetization, 103
electric displacement, 86, 91–93, 95–97, 112, 113
magnetic excitation, 86, 103–105, 107, 109, 110, 114
polarization, 86, 87, 90, 92, 96
magnetization, 98, 106, 110
Poynting, 66, 68, 69, 127

W

wave
  complex, 84
  electromagnetic, 69, 85, 119, 120, 124–126, 129, 131, 139, 148, 167, 179
  incident, 73
  matter, 72, 73
  number, 125, 221
  phase, 125, 126
  plane, 69, 72–78, 80–82, 84
  reflected, 73, 80, 81
  stationary, 83, 84
wavelength, 126, 132, 133, 135, 151, 160, 163, 168, 169, 176, 209–217, 221, 232, 261–263
weber (Wb), 29
Worldwide Interoperability for Microwaves Access (WIMAX), 147, 172, 173